TRAINING FOR CERTIF...

Taking An ASE Certification Test

This study guide will help prepare you to take and pass the ASE test. It contains descriptions of the types of questions used on the test, the task list from which the test questions are derived, a review of the task list subject information, and a practice test containing ASE style questions.

ABOUT ASE

The National Institute for Automotive Service Excellence (ASE) is a non-profit organization founded in 1972 for the purpose of improving the quality of automotive service and repair through the voluntary testing and certification of automotive technicians. Currently, there are over 400,000 professional technicians certified by ASE in over 40 different specialist areas.

ASE certification recognizes your knowledge and experience, and since it is voluntary, taking and passing an ASE certification test also demonstrates to employers and customers your commitment to your profession. It can mean better compensation and increased employment opportunities as well.

ASE not only certifies technician competency, it also promotes the benefits of technician certification to the motoring public. Repair shops that employ at least one ASE technician can display the ASE sign. Establishments where 75% of technicians are certified, with at least one technician certified in each area of service offered by the business, are eligible for the ASE Blue Seal of Excellence program. ASE encourages consumers to patronize these shops through media campaigns and car care clinics.

To become ASE certified, you must pass at least one ASE exam and have at least two years of related work experience. Technicians that pass all tests in a series earn Master Technician status. Your certification is valid for five years, after which time you must retest to retain certification, demonstrating that you have kept up with the changing technology in the field.

THE ASE TEST

An ASE test consists of forty to eighty multiple-choice questions. Test questions are written by a panel of technical experts from vehicle, parts and equipment manufacturers, as well as working technicians and technical education instructors. All questions have been pre-tested and quality checked on a national sample of technicians. The questions are derived from information presented in the task list, which details the knowledge that a technician must have to pass an ASE test and be recognized as competent in that category. The task list is periodically updated by ASE in response to changes in vehicle technology and repair techniques.

© Advanstar 2004

Customer Service 1-800-240-1968
FAX 218-723-9437
e-mail www.ma@superfill.com
URL: www.motorage.com

Taking An ASE Certification Test

There are five types of questions on an ASE test:

Direct, or Completion
MOST Likely
Technician A and Technician B
EXCEPT
LEAST Likely

Direct, or Completion

This type of question is the kind that is most familiar to anyone who has taken a multiple-choice test: you must answer a direct question or complete a statement with the correct answer. There are four choices given as potential answers, but only one is correct. Sometimes the correct answer to one of these questions is clear, however in other cases more than one answer may seem to be correct. In that case, read the question carefully and choose the answer that is most correct. Here is an example of this type of test question:

A compression test shows that one cylinder is too low. A leakage test on that cylinder shows that there is excessive leakage. During the test, air could be heard coming from the tailpipe. Which of the following could be the cause?
  A. broken piston rings
  B. bad head gasket
  C. bad exhaust gasket
  D. an exhaust valve not seating

There is only one correct answer to this question, answer **D**. If an exhaust valve is not seated, air will leak from the combustion chamber by way of the valve out to the tailpipe and make an audible sound. Answer C is wrong because an exhaust gasket has nothing to do with combustion chamber sealing. Answers A and B are wrong because broken rings or a bad head gasket would have air leaking through the oil filler or coolant system.

MOST Likely

This type of question is similar to a direct question but it can be more challenging because all or some of the answers may be nearly correct. However, only one answer is the most correct. For example:

When a cylinder head with an overhead camshaft is discovered to be warped, which of the following is the **MOST** correct repair option?
  A. replace the head
  B. check for cracks, straighten the head, surface the head
  C. surface the head, then straighten it
  D. straighten the head, surface the head, check for cracks

The most correct answer is **B**. It makes no sense to perform repairs on a cylinder head that might not be useable. The head should first be checked for warpage and cracks. Therefore, answer B is more correct than answer D. The head could certainly be replaced, but the cost factor may be prohibitive and availability may be limited, so answer B is more correct than answer A. If the top of the head is warped enough to interfere with cam bore alignment and/or restrict free movement of the camshaft, the head must be straightened before it is resurfaced, so answer C is wrong.

Technician A and Technician B

These questions are the kind most commonly associated with the ASE test. With these questions you are asked to choose which technician statement is correct, or whether they both are correct or incorrect. This type of question can be difficult because very often you may find one technician's statement to be clearly correct or incorrect while the other may not be so obvious. Do you choose one technician or both? The key to answering these questions is to carefully examine each technician's statement independently and judge it on its own merit. Here is an example of this type of question.

A vehicle equipped with rack-and-pinion steering is having the front end inspected. Technician A says that the inner tie rod ends should be inspected while in their normal running position. Technician B says that if movement is felt between the tie rod stud and the socket while the tire is moved in and out, the inner tie rod should be replaced. Who is correct?
  A. Technician A
  B. Technician B
  C. Both A and B
  D. Neither A or B

The correct answer is **C**; both technicians' statements are correct. Technician B is clearly correct because any play felt between the tie-rod stud and the socket while the tire is moved in and out indicates that the assembly is worn and requires replacement. However, Technician A is also correct because inner tie-rods should be inspected while in their normal running position, to prevent binding that may occur when the suspension is allowed to hang free.

EXCEPT

This kind of question is sometimes called a negative question because you are asked to give the incorrect answer. All of the possible answers given are correct EXCEPT one. In effect, the correct answer to the question is the one that is wrong. The word EXCEPT is always capitalized in these questions. For example:

All of the following are true of torsion bars **EXCEPT:**
    A. They can be mounted longitudinally or transversely.
    B. They serve the same function as coil springs.
    C. They are interchangeable from side-to-side.
    D. They can be used to adjust vehicle ride height.

The correct answer is **C**. Torsion bars are not normally interchangeable from side-to-side. This is because the direction of the twisting or torsion is not the same on the left and right sides. All of the other answers contain true statements regarding torsion bars.

LEAST Likely

This type of question is similar to EXCEPT in that once again you are asked to give the answer that is wrong. For example:

Blue-gray smoke comes from the exhaust of a vehicle during deceleration. Of the following, which cause is **LEAST** likely?
    A. worn valve guides
    B. broken valve seals
    C. worn piston rings
    D. clogged oil return passages

The correct answer is **C**. Worn piston rings will usually make an engine smoke worse under acceleration. All of the other causes can allow oil to be drawn through the valve guides under the high intake vacuum that occurs during deceleration.

PREPARING FOR THE ASE TEST

Begin preparing for the test by reading the task list. The task list describes the actual work performed by a technician in a particular specialty area. Each question on an ASE test is derived from a task or set of tasks in the list. Familiarizing yourself with the task list will help you to concentrate on the areas where you need to study.

The text section of this study guide contains information pertaining to each of the tasks in the task list. Reviewing this information will prepare you to take the practice test.

Take the practice test and compare your answers with the correct answer explanations. If you get an answer wrong and don't understand why, go back and read the information pertaining to that question in the text.

After reviewing the tasks and the subject information and taking the practice test, you should be prepared to take the ASE test or be aware of areas where further study is needed. When studying with this study guide or any other source of information, use the following guidelines to make sure the time spent is as productive as possible:

- Concentrate on the subject areas where you are weakest.
- Arrange your schedule to allow specific times for studying.
- Study in an area where you will not be distracted.
- Don't try to study after a full meal or when you are tired.
- Don't wait until the last minute and try to 'cram' for the test.

TAKING THE ASE TEST

Make sure you get a good night's sleep the night before the test. Have a good lunch on test day but either eat lightly or skip dinner until after the test. A heavy meal will make you tired.

Bring your admission ticket, some form of photo identification, three or four sharpened #2 pencils and a watch (to keep track of time as the test room may not have a clock) with you to the test center.

The test proctor will explain how to fill out the answer sheet and how much time is allotted for each test. You may take up to four certification tests in one sitting, but this may prove too difficult unless you are very familiar with the subject areas.

When the test begins, open the test booklet to see how many questions are on the test. This will help you keep track of your progress against the time remaining. Mark your answer sheet clearly, making sure the answer number and question number correspond.

Read through each question carefully. If you don't know the answer to a question and need to think about it, move on to the next question. Don't spend too much time on any one question. After you have worked through to the end of the test, check your remaining time and go back and answer the questions you had

Taking An ASE Certification Test

trouble with. Very often, information found in questions later in the test can help answer some of the ones with which you had difficulty.

If you are running out of time and still have unanswered test questions, guess the answers if necessary to make sure every question is answered. Do not leave any answers blank. It is to your advantage to answer every question, because your test score is based on the number of correct answers. A guessed answer could be correct, but a blank answer can never be.

To learn exactly where and when the ASE Certification Tests are available in your area, as well as the costs involved in becoming ASE certified, please contact ASE directly for a registration booklet.

The National Institute for
Automotive Service Excellence
101 Blue Seal Drive, S.E.
Suite 101
Leesburg, VA 20175

1-877-ASE-TECH (273-8324)

http://www.asecert.org

Engine Repair

TEST SPECIFICATIONS
FOR ENGINE REPAIR (TEST A1)

CONTENT AREA	NUMBER OF QUESTIONS IN ASE TEST	PERCENTAGE OF COVERAGE IN ASE TEST
A. General Engine Diagnosis	17	28
B. Cylinder Head And Valvetrain Diagnosis And Repair	14	23
C. Engine Block Diagnosis And Repair	14	23
D. Lubrication And Cooling Systems Diagnosis And Repair	8	13
E. Fuel And Exhaust Systems Inspection And Service	7	12
Total	60	100%

The 5-year Recertification Test will cover the same content areas as those listed above. However, the number of questions in each content area of the Recertification Test will be reduced by about one-half.

The following pages list the tasks covered in each content area. These task descriptions offer detailed information to technicians preparing for the test, and to persons who may be instructing technicians in Engine Repair. The task list may also serve as a guideline for question writers, reviewers and test assemblers.

It should be noted that the number of questions in each content area may not equal the number of tasks listed. Some of the tasks are complex and broad in scope, and may be covered by several questions. Other tasks are simple and narrow in scope; one question may cover several tasks. The main purpose for listing the tasks is to describe accurately what is done on the job, not to make each task correspond to a particular test question.

ENGINE REPAIR TEST TASK LIST

A. GENERAL ENGINE DIAGNOSIS
(17 questions)

Task 1 - Verify driver's complaint and/or road test vehicle; determine necessary action.
Task 2 - Determine if no-crank, no-start or hard starting condition is an ignition system, cranking system, fuel system, exhaust system or engine mechanical problem.
Task 3 - Inspect engine assembly for fuel, oil, coolant and other leaks; determine necessary action.
Task 4 - Listen to engine noises and vibrations; determine necessary action.
Task 5 - Diagnose the cause of excessive oil consumption, coolant consumption, unusual engine exhaust color, odor and sound; determine necessary action.
Task 6 - Perform engine vacuum tests; determine necessary action.
Task 7 - Perform cylinder power balance tests; determine necessary action.
Task 8 - Perform cylinder cranking compression tests; determine necessary action.
Task 9 - Perform cylinder leakage tests; determine necessary action.

B. CYLINDER HEAD AND VALVETRAIN DIAGNOSIS AND REPAIR
(14 questions)

Task 1 - Remove cylinder heads, disassemble, clean and prepare for inspection.
Task 2 - Visually inspect cylinder heads for cracks, warpage, corrosion and leakage, and check passage condition; determine needed repairs.
Task 3 - Inspect and verify valve springs for squareness, pressure and free height comparison; replace as necessary.
Task 4 - Inspect valve spring retainers, rotators, locks and valve lock grooves.
Task 5 - Replace valve stem seals.
Task 6 - Inspect valve guides for wear; check valve guide height and stem-to-guide clearance; determine needed repairs.
Task 7 - Inspect valves and valve seats; determine needed repairs.
Task 8 - Check valve face-to-seat contact and valve seat concentricity (runout).
Task 9 - Check valve spring installed (assembled) height and valve stem

height; determine needed repairs.

Task 10 - Inspect pushrods, rocker arms, rocker arm pivots and shafts for wear, bending, cracks, looseness and blocked oil passages; repair or replace as required.

Task 11 - Inspect and replace hydraulic or mechanical lifters/lash adjusters.

Task 12 - Adjust valves on engines with mechanical or hydraulic lifters.

Task 13 - Inspect and replace camshaft(s) (includes checking gear wear and backlash, end-play, sprocket and chain wear, overhead cam drive sprocket(s), drive belt(s), belt tension, tensioners and cam sensor components).

Task 14 - Inspect and measure camshaft journals and lobes; measure camshaft lift.

Task 15 - Inspect and measure camshaft bore for wear, damage, out-of-round and alignment; determine needed repairs.

Task 16 - Time camshaft(s) to crankshaft.

Task 17 - Inspect cylinder head mating surface condition and finish, reassemble and install gasket(s) and cylinder head(s); replace and tighten fasteners according to manufacturers' procedures.

C. ENGINE BLOCK DIAGNOSIS AND REPAIR
(14 questions)

Task 1 - Disassemble engine block; clean and prepare components for inspection and reassembly.

Task 2 - Visually inspect engine block for cracks, corrosion, passage condition, core and gallery plug hole condition, surface warpage and surface finish and condition; determine necessary action.

Task 3 - Inspect and repair damaged threads where allowed; install core and gallery plugs.

Task 4 - Inspect and measure cylinder walls; remove cylinder wall ridges; hone and clean cylinder walls; determine need for further action.

Task 5 - Inspect crankshaft for end-play, straightness, journal damage, keyway damage, thrust flange and sealing surface condition and visual surface cracks; check oil passage condition; measure journal wear; check crankshaft sensor reluctor ring (where applicable); determine necessary action.

Task 6 - Inspect and measure main bearing bores and cap alignment and fit.

Task 7 - Install main bearings and crankshaft; check bearing clearances and end-play; replace/retorque bolts according to manufacturers' procedures.

Task 8 - Inspect camshaft bearings for excessive wear and alignment; install camshaft, timing chain and gears; check end-play.

Task 9 - Inspect auxiliary (balance, intermediate, idler, counterbalance or silencer) shaft(s), drive(s) and support bearings for damage and wear; determine necessary action.

Task 10 - Inspect, measure, service or replace pistons and piston pins; identify piston and bearing wear patterns that indicate connecting rod alignment problems; determine necessary action.

Task 11 - Inspect connecting rods for damage, bore condition and pin fit; determine necessary action.

Task 12 - Inspect, measure and install or replace piston rings; assemble piston and connecting rod; install piston/rod assembly; check bearing clearance and sideplay; replace/retorque fasteners according to manufacturers' procedures.

Task 13 - Inspect, reinstall or replace crankshaft vibration damper (harmonic balancer).

Task 14 - Inspect crankshaft flange and flywheel mating surfaces; inspect and replace crankshaft pilot bearing/bushing (if applicable); inspect flywheel/flexplate for cracks and wear (including flywheel ring gear); measure flywheel runout; determine necessary action.

Task 15 - Inspect and replace pans and covers.

Task 16 - Assemble the engine using gaskets, seals and formed-in-place (tube-applied) sealants, thread sealers, etc. according to manufacturers' specifications.

D. LUBRICATION AND COOLING SYSTEMS DIAGNOSIS AND REPAIR
(8 questions)

Task 1 - Diagnose engine lubrication system problems; perform oil pressure tests; determine necessary action.

Task 2 - Disassemble, inspect and measure oil pump (includes gears, rotors, housing and pick-up assembly), pressure relief devices and pump drive; determine necessary action; replace oil filter.

Task 3 - Perform cooling system tests; determine necessary action.

Task 4 - Inspect, replace and adjust drive belt(s), tensioner(s) and pulleys.

Task 5 - Inspect and replace engine cooling and heater system hoses and fittings.

Task 6 - Inspect, test and replace thermostat, bypass and housing.

Task 7 - Inspect coolant; drain, flush and refill cooling system with recommended coolant; bleed air as required.

Task 8 - Inspect and replace water pump.

Task 9 - Inspect, test and replace radiator, heater core, pressure cap and coolant recovery system.

Task 10 - Clean, inspect, test and replace fan (both electrical and mechanical), fan clutch, fan shroud, air dams and cooling-related temperature sensors.

Task 11 - Inspect, test and replace internal and external oil coolers.

E. FUEL, ELECTRICAL, IGNITION AND EXHAUST SYSTEMS INSPECTION AND SERVICE
(7 questions)

Task 1 - Inspect, clean or replace

fuel and air induction system components, intake manifold and gaskets.

Task 2 - Inspect, service or replace air filters, filter housings and intake ductwork.

Task 3 - Inspect turbocharger/supercharger; determine necessary action.

Task 4 - Test engine cranking system; determine needed repairs.

Task 5 - Inspect and replace positive crankcase ventilation (PCV) system components.

Task 6 - Visually inspect and reinstall primary and secondary ignition system components; time distributor.

Task 7 - Inspect and diagnose exhaust system; determine needed repairs.

The preceding Task List Data details all of the related informational subject matter you are expected to know in order to sit for this ASE Certification Test. Your own years of experience in the professional automotive service trade as a technician also should provide you with added background.

Finally, a conscientious review of the self-study material provided in this Training for ASE Certification unit also should help you to be adequately prepared to take this test.

NOTES

TRAINING FOR CERTIFICATION

GENERAL ENGINE DIAGNOSIS

Any engine repair should begin with correct diagnosis, which may in fact reveal that a mechanical repair is not required. It would be unfortunate to perform a significant mechanical repair like an engine overhaul, only to determine that the real malfunction was in another system.

Also, make sure you diagnose and repair the real problem and not just focus on the results of the problem. In many cases something other than the engine malfunctioned and in turn caused the engine to fail. This problem must be addressed before the vehicle leaves or the engine will only fail again. It doesn't make any sense to overhaul an engine that overheated without repairing the cause of the cooling system failure.

Proper engine diagnosis begins with listening to the customer's complaint. Then, visually inspect the engine assembly and listen for abnormal engine noise. If the engine is using excessive amounts of oil or coolant, determine the cause and check the exhaust color and odor. Finally, perform vacuum, compression, cylinder leakage and cylinder balance tests in order to determine your course of action.

The first step in diagnosis is talking to the customer. Get a detailed description of what happens and under what conditions it happens, then verify the information. Conduct a road test if possible. Before road testing the vehicle, visually inspect the engine for obvious problems such as:
• fuel, oil and coolant leaks
• loose, cracked, glazed or frayed belts
• broken, burned or damaged wiring and loose or corroded connectors
• soft, brittle, kinked or damaged hoses
• loose components and missing fasteners
• clogged air filter and missing or damaged air cleaner duct work.

If the engine won't start or is hard to start, you'll have to determine if the reason is due to a cranking system, ignition system, fuel system or engine mechanical problem.

If the engine won't crank, check the battery terminal connections and make sure they are clean and tight. If the battery has removable caps, check the electrolyte level and specific gravity. On sealed maintenance-free batteries, check the color of the built-in charge indicator. Check the open circuit voltage using a voltmeter and charge the battery if necessary.

If the battery is good and the engine still won't crank, check the cables, wiring and connections in the starter circuit. Make sure the solenoid is functioning and the starter is properly mounted. Perform a starter current draw test. Current draw that is greater than specification could be caused by a bad starter or a binding mechanical assembly. Remove the spark plugs. If the engine can be turned by hand using suitable tools at the flywheel or crankshaft balancer, then the problem is with the starter. If the engine cannot be turned by hand, it may be seized or have broken internal components.

If the engine cranks, but seems to crank too easily, the timing belt or chain may have broken or jumped time. Rotate the crankshaft by hand while watching for distributor rotor or camshaft movement. If the rotor or camshaft do not move when the crankshaft is turned, the belt or chain is broken. If they turn when the crankshaft is turned, bring the piston in cylinder No. 1 to TDC (Top Dead Center) on the compression stroke and verify that the valve timing is correct. A broken timing belt or timing chain, or incorrect valve timing caused by a belt or chain that has jumped time, can result in valve to piston contact on some engines.

If the cam drive system and valve timing are OK and the engine cranks but won't start, check for spark and adequate fuel delivery. Check for spark at the spark plug wires using a spark tester. If the spark plugs do not fire, first determine in which part of the ignition circuit the fault lies. Disconnect the coil wire from the distributor cap and check for a good spark with the spark tester while cranking the engine. If there is a good spark, the problem is in the ignition sec-

ondary circuit (distributor cap, rotor, wires and plugs). If there is no spark or a weak spark, the problem is in the ignition primary circuit. Check the ignition module and coil. If they are OK, continue testing components, wiring and connections back through the primary circuit.

Check the fuel delivery system by disconnecting the fuel pressure feed line and inserting it into a graduated container. On systems with mechanical fuel pumps, crank the starter motor. On systems with electric fuel pumps, you usually must operate the pump with a jumper wire for a specified period. First, you should be able to hear the electric fuel pump running in the tank. If not, the pump must be removed for further inspection. If the electric pump runs, it should flow at least a half-pint of fuel in 30 seconds. If not, before condemning the pump, check the external and in-tank fuel filters and the fuel lines for restriction. Also, perform a voltage drop test on the power and ground circuits to the pump.

At this point, the engine should be cranking properly, have spark and adequate fuel delivery. If the engine still does not start, further diagnosis of the fuel and electronic engine control system are needed.

ENGINE LEAKS

Fuel Leaks

Since the fuel system is under pressure when the engine is running, the source of a fuel leak is usually not difficult to find. Fuel can leak from damaged lines or hoses or from loose fittings. Inspect hoses for cracks and swelling and inspect the condition and security of hose clamps. Inspect fuel lines for cracks, corrosion and damage from abrasion. Make sure all fittings are properly installed and tightened. A leak from a fitting that is tight may be caused by a damaged O-ring. On vehicles with fuel injection, components like the fuel pressure regulator can also be the source of a fuel leak.

Oil Leaks

Oil leaks are most often caused by hardened or worn out seals and gaskets, or leaking oil pressure sending units. However, engine oil leaks can also be caused by something technicians often overlook—excessive crankcase pressure, which can be caused by worn rings or excessive cylinder wall clearances. These problems allow an unusual amount of combustion blow-by gases to enter the crankcase, where the gases can push oil past seals and gaskets that are in good condition.

The key here is to look for related symptoms, such as spark plugs that are oil-fouled or show signs of oil deposits. Look for the telltale blue-gray or gray-white exhaust smoke of an oil-burner.

Excessive crankcase pressure can also build up in a perfectly good engine if the crankcase ventilation system isn't working properly. Remember that fresh air must be able to enter the crankcase. This means that the crankcase breather filter must be clean and the related plumbing must be unobstructed. It also means that the PCV valve must work correctly and its plumbing must be clean and in good condition.

At idle, a substantial amount of air enters the PCV valve. If you pull the PCV valve out of its grommet with the engine idling, you should feel a strong suction at the valve's inlet. Furthermore, the idle speed should drop at least 50 rpm when you cover the inlet of the PCV valve.

Many systems use a PCV orifice in place of the PCV valve. If the oil is not changed on a regular basis, this system will plug up. Some systems may have screens that can be removed and cleaned.

Sometimes you'll see oil on the ground under the engine, smell it burning off of hot engine parts, and see it dripping from the engine but still not be able to tell exactly where it's coming from. For hard to find oil leaks, add a fluorescent dye to the engine oil that is visible with a black light. Run the engine for a while and then pass the light around the engine. The dye should pinpoint the source of the leak.

Coolant Leaks

Visually inspect for coolant leaks at the radiator and heater hoses, water pump, radiator, intake manifold, sensor fittings, water control valves and heater core. Attach a suitable pressure tester to the coolant filler neck and apply pressure equal to the pressure rating marked on the radiator cap (make sure the cap is the correct one for the vehicle!). The pressure should remain at that level if there are no leaks in the cooling system. If the pressure drops, check for leaks in the same areas.

External coolant leaks are relatively easy to find. But, what if you pressurize the cooling system with a cooling system pressure tester, the pressure drops, and no coolant appears outside the engine? This indicates that the coolant leak is inside the engine and is most likely caused by a cracked cylinder head, cracked block or blown head gasket.

Symptoms will vary depending upon the severity of an internal coolant leak. When coolant leaks into a cylinder, it may create white exhaust smoke and a somewhat sweet antifreeze odor in the exhaust. It may also cause misfiring, especially when the engine is cold. Unfortunately,

TRAINING FOR CERTIFICATION

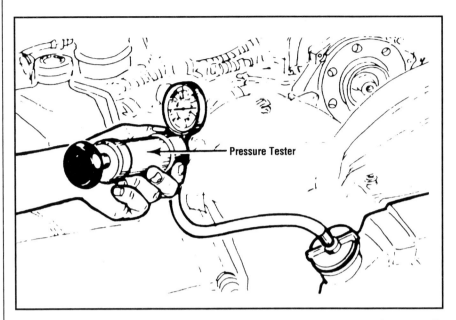

Pressure testing the cooling system. Watch the gauge needle for an indication of a cooling system leak.

catalytic converters can mask small coolant leak symptoms. The converter super-heats the coolant into such a fine vapor that it is not noticeable. A coolant leak can also damage oxygen sensors. If a coolant leak has been confirmed and repaired, always test the oxygen sensor for proper operation, or driveability symptoms may remain.

If coolant is leaking into a cylinder, combustion gases will also be able to escape into the cooling system. When combustion gas escapes into the cooling system, it can cause big air bubbles to appear in the radiator coolant when the engine is running. It can also pressurize the coolant recovery reservoir.

One common internal coolant leak detection procedure uses a chemical that is sensitive to combustion gas. With the engine running, place a vial of the chemical over the radiator neck and draw vapors from the top of the radiator into the vial. If the chemical changes color, you know that combustion gases are leaking into the cooling system.

Another way to detect combustion gas is to carefully hold the exhaust analyzer probe over the neck of the radiator. Do not allow the probe to draw in coolant. Cup a clean cloth or a plastic bag with a hole for the probe over the radiator neck. If combustion gas is entering the coolant, you'll see a reading on the exhaust analyzer.

Sometimes when you disable cylinders, the bubbles appearing in the radiator will diminish when you eliminate the cylinder that has the coolant leak. Disabling the leaking cylinder may also reveal little or no rpm drop compared to the other cylinders. Remember, in closed loop the computer may adjust idle and mixture faster than you can hear or see a rpm change.

Closely examine the spark plugs. When combustion heat evaporates the leaking coolant, it not only creates white exhaust smoke—it also tends to clean the porcelain insulator that surrounds the plug's center electrode.

If the radiator is cool enough to do so, remove the radiator cap. Remove all the spark plugs. Then perform a leakdown test on the cylinder with the unusually clean-looking spark plug. If pumping air into that cylinder produces bubbles in the radiator coolant, you know that the leak is in that cylinder. If you also find that air escapes from an adjacent cylinder, you know that the head gasket has blown out between those two cylinders and has caused both a compression leak and a coolant leak.

In cases where disabling cylinders or inspecting the spark plugs fail to point to a particular cylinder, perform a leakdown test on each cylinder. After you have pinpointed the leaking cylinder and have disassembled the engine, always inspect both the block and the head for cracks and warpage.

ENGINE NOISES

Generally, noises are caused by too much clearance between parts or loss of oil pressure. The following common engine noises can be caused by any of these parts:

Crankshaft Noises
- Main bearings
- Connecting rod bearings
- Pistons
- Wrist pin
- Crankshaft end-play.

Valvetrain Noises
- Bearing noise
- Rocker arms, shaft, ball and seat
- Pushrods
- Tappets and camshaft
- Timing gears and chain.

Other Noises
- Loose or broken brackets
- Oil pump failure
- Spark knock.

In order to successfully diagnose noises, you must pay close attention to the frequency at which the noise occurs and how the frequency changes as you vary engine load and engine rpm. Also note how factors such as temperature and oil pressure affect the noise.

Bearing Noises

Main bearing noise is caused by too much bearing clearance, which usually creates a dull or deep-sounding metallic knocking. When you increase rpm or engine load, the knocking usually increases in frequency. The noise is usually most obvious right after the engine starts up, when the engine is under a heavy load, or during acceleration. Along with the knocking sound, the engine may also exhibit low oil pressure.

Connecting rod noise, which is also caused by excessive bearing clearance, is much less intense than main bearing noise. This noise usually sounds like a light metallic rapping that is most noticeable when the engine is running under a light load at relatively slow speeds. This knock becomes louder and occurs more frequently when the speed of the engine is increased. When you eliminate the ignition or injection to the cylinder with a rod knock, the sound diminishes.

Crankshaft End-Play Noise

When crankshaft end-play is excessive, the engine may make a sharp, irregular metallic rapping sound. The noise is usually most obvious at idle and becomes louder when you engage and release the clutch on a manual transmission-equipped vehicle.

Where space allows, you can verify excessive crankshaft end-play by fitting a dial indicator to the tip of the crankshaft. Using a prybar, carefully pry the crankshaft back-and-forth and note the reading on the dial indicator. Compare the reading with specifications.

Piston Noises

Excessive piston-to-wall clearance can cause piston slap. This is caused by side-to-side movement of the piston within the cylinder bore, and sounds like a dull or muffled metallic rattle at idle or during light engine loads.

Very faint piston slap may disappear after the engine warms up and the piston expands. In this case, the piston-to-wall clearance usually isn't severe enough to worry about. However, piston slap that continues after the engine warms up should be corrected. Note that unlike a connecting rod bearing noise, piston slap does not quiet down and may in fact grow louder when you eliminate ignition or fuel injection to that cylinder.

A knocking noise can be caused by excessive carbon buildup in the combustion chamber where the piston contacts the carbon at TDC.

Piston Pin Noise

When piston-to-piston pin clearance is excessive, the pin makes a light but sharp metallic rapping at idle. The sound may be more obvious during low-speed driving. Eliminating ignition or fuel injection to a cylinder with a loose piston pin will change the frequency and possibly the intensity of the rapping noise.

Hydraulic Lifter Noise

A noisy hydraulic lifter is usually a consistent ticking sound that occurs at a frequency slower than engine rpm (remember that the valvetrain operates at half crankshaft rpm). Try sliding a feeler gauge between the valve stem and the rocker arm. If this eliminates the ticking, it confirms that there is excessive clearance in the valvetrain.

With the engine running, you can also press down on the pushrod end of each rocker arm with a piece of wood or a hammer handle. If this stops or reduces the ticking, you have pinpointed the faulty lifter. Always check valve adjustment and inspect valvetrain parts for wear or damage. Worn valvetrain parts can mimic the noise of bad hydraulic lifters.

Spark Knock

Spark knock, which is caused by two kinds of abnormal, uncontrolled combustion, sounds like a metallic pinging or ringing noise. You may hear spark knock when the engine is under a heavy load and being run at too low an rpm, or when the engine is accelerating. An engine that is running too hot and/or has excessive combustion deposits can also suffer from spark knock.

Preignition spark knock occurs when a hot piece of carbon or metal inside the combustion chamber prematurely ignites the air/fuel mixture. Then the spark plug ignites the remaining mixture at the normal time. When the two portions of burning mixture meet each other inside the combustion chamber, there is a sudden and abnormal rise in cylinder pressure, which causes engine vibration in that cylinder.

Detonation spark knock is primarily caused by fuel with too low an octane rating for the engine, ignition timing that is too far advanced, high engine operating temperature or excessive carbon buildup in the combustion chamber. If the octane rating is too low for the engine, it basically means that the fuel will burn too quickly. When detonation occurs, the spark fires, the mixture begins burning, and pressure in the cylinder begins rising. But the rise in normal pressure causes part of the air/fuel mixture elsewhere in the combustion chamber to self-ignite. Then the two flame fronts collide as in the preignition situation.

Both preignition and detonation can cause damage to pistons and spark plugs.

TRAINING FOR CERTIFICATION

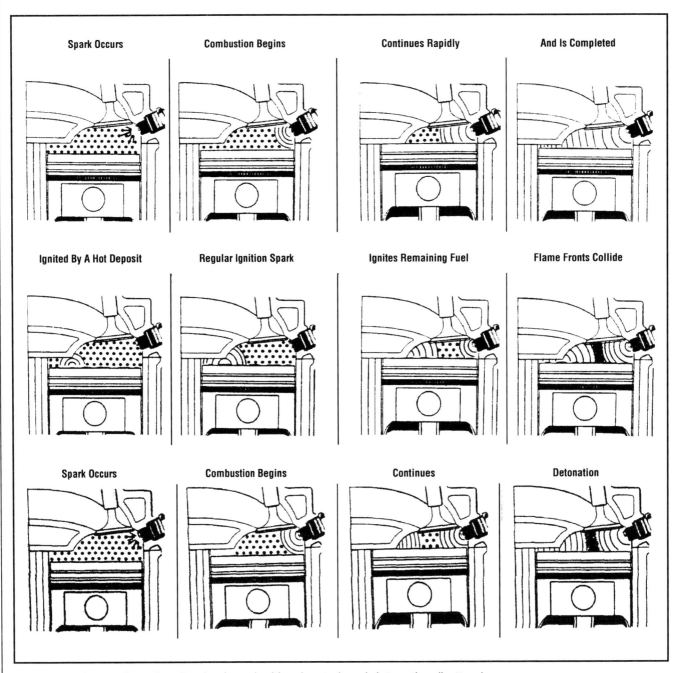

Examples of normal combustion (top), preignition (center) and detonation (bottom).

Most vehicles have a knock sensor to adjust timing as needed to avoid damage, however, excessive spark knock may be a sign of engine control problems. Before any major engine tear down, check the engine control system. EGR operation, knock control, coolant level or any number of engine sensors or solenoids can cause spark knock. Asking the customer how long the engine has made this noise is the best way to determine how much damage may have occurred.

Other Noises

A knock that is most noticeable when the vehicle is in Park or Neutral can be caused by loose torque converter bolts.

A tapping sound that may sound like a valvetrain noise can actually be caused by an exhaust leak at the exhaust manifold/cylinder head juncture.

EXCESSIVE OIL CONSUMPTION AND EXHAUST SMOKE

Diagnosing Exhaust Smoke

Black exhaust smoke indicates

that the air/fuel mixture is too rich. Black smoke is usually caused by a defective fuel injector or electronic sensor misleading the computer to add more fuel than is needed. It is rare that black smoke is engine mechanical-related.

Excessive white exhaust smoke usually means coolant is leaking into one or more cylinders and the engine is trying to burn off that coolant. The most common cause of white smoke is a bad head gasket. A compression test must be performed to narrow down the damaged area. Any leak from the cooling system into the intake or combustion areas can cause white smoke. This problem may be masked until it becomes severe, because the catalytic converter will super-heat the water to such a fine vapor that it may not be noticeable.

Blue-gray or gray-white smoke tells you the engine is burning oil. This could be caused by something as simple as a PCV system malfunction. The most common causes of burning oil are worn valve guides and/or seals, and worn piston rings and/or cylinders. A cylinder leakage test must be performed to determine if the problem is located in the cylinder head or block.

Abnormal Oil Consumption

The three common causes of excessive oil consumption are oil leaks, valve guide/valve seal problems, and piston ring problems.

When an engine is using oil, inspect it visually for serious oil leaks first. Check the crankcase ventilation system as a matter of routine. If you don't, the oil leaks may reappear in spite of the new seals and gaskets you install.

A careful road test can be a critical diagnostic step in determining the cause of the oil burning. Note when the oil smoke is most intense. Typically, bad piston rings will make the engine smoke worse when it is accelerating—especially after it has been idling for a long time.

When a vehicle suffers from worn valve guides and/or bad valve stem seals, you'll see exhaust smoke during deceleration. The high intake vacuum that occurs during deceleration draws the oil through the worn guides or seals. Remember, the catalytic converter will super-heat the oil and reduce some of the smoke that would have been seen on pre-converter models.

Reading the spark plugs can also confirm your diagnosis. When the excess oil is coming from the valve guide, the oil deposits on the spark plug tend to accumulate only on one side of the spark plug. When the excess oil is coming from the rings, the deposits on the spark plug tend to accumulate around the entire spark plug.

Before you blame either the valve guides or the valve seals for an oil consumption problem, verify that all of the oil return holes are clean. If oil cannot drain freely back into the crankcase, it can accumulate in the head and be drawn by vacuum into the combustion chamber, increasing oil consumption and causing exhaust smoke. Valve stem seals are designed to keep normal amounts of lubricating oil from entering the combustion chamber, but seldom work well when submerged in oil.

ENGINE VACUUM TESTS

Cranking Vacuum Test

Properly disable the ignition and fuel systems. Connect an ammeter to the starting circuit and a vacuum gauge to the intake manifold. Crank the engine, listen to the cranking rhythm, and watch your instruments.

On a good engine, the cranking speed and cranking rhythm will sound crisp and consistent. There will be no 'pauses' or uneven rhythms, suggesting binding or differing compression values. The ammeter will stabilize at a consistent current draw reading that's within specifications. The vacuum gauge will read a fairly steady 3 to 5 inches Hg or more.

The better the rings and valves are sealing, the stronger cranking vacuum will be. All things being equal, the following are true:

1. The stronger the cranking vacuum, the quicker the engine will start.

2. The weaker the cranking vacuum, the more difficult it will be to start the engine. Moreover, if the engine can't draw any cranking vacuum at all, it won't start.

Whenever you see zero or nearly zero cranking vacuum, check for a substantial air leak such as:
- an improperly adjusted throttle blade
- a loose or cracked carburetor or intake throttle body
- a stuck-open PCV valve or a cracked PCV hose
- secondary throttle blades that are stuck open (where a carburetor is used)
- a leaking intake manifold gasket.

Zero cranking vacuum and a no-start or hard-start complaint can also be caused by a severe exhaust restriction. When in doubt, loosen the exhaust pipe(s) at the exhaust manifold(s) and repeat the test. An easier way to test quickly on many engines is to remove the oxygen sensor. Strong puffs of air from the mounting hole during cranking can indicate a restricted exhaust. A simple gauge is available to screw into the hole to measure the backpressure.

Of course, poor cranking vacuum (coupled with faster-than-normal cranking speed) could also mean that compression is low in all cylinders due to normal

15

TRAINING FOR CERTIFICATION

engine wear or due to a valve timing or timing belt problem. When timing chains or timing belts wear or stretch, valve timing can go astray. Sometimes, the camshaft drive system will literally jump a tooth and the engine will continue running (although very sluggishly).

Note that with some jumped valve timing problems, the engine will crank very unevenly and the vacuum and ammeter readings will be very erratic. You may notice that disabling the ignition on an erratically cranking engine makes the engine crank smoothly again. This indicates an ignition or valve timing problem.

Suppose the engine has one or more consistent compression leaks. Every time the cylinder with the compression leak comes around:
• the compression air volume will drop momentarily
• the cranking speed will increase momentarily
• the starter current draw will decrease momentarily.

The reason that the cranking speed increases and starter draw decreases is that it takes less effort for the starter to crank a weaker cylinder. When the cranking tests suggest a compression problem, you must perform other pinpoint tests to confirm the source of the problem. Start with compression and cylinder leakdown tests.

Manifold Vacuum Test

Checking manifold vacuum can reveal a variety of engine maladies. When the engine has reached normal operating temperature, connect a vacuum gauge to a manifold vacuum port. As a general rule, an engine in good condition should produce a steady 17 to 21-in. Hg reading at idle. However, always check the standard for the particular engine in question.

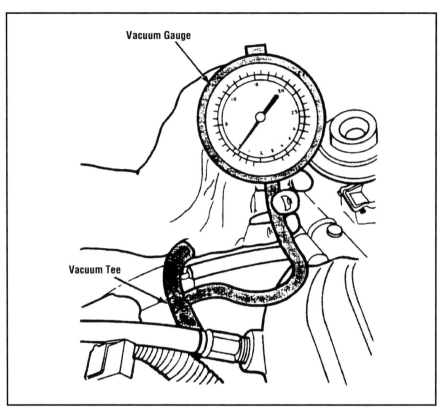

Typical vacuum gauge installed for manifold vacuum test. *(Courtesy: Ford Motor Co.)*

NOTE: *Remember atmospheric pressure changes with elevation. Manufacturers provide sea level readings so the technician needs to adjust readings accordingly. As an approximation, for every 1000 feet above sea level, remove one inch of vacuum.*

Does the idle vacuum look OK? If so, disconnect the vacuum hose from the EGR valve and plug it. Using normal safety precautions, slowly raise engine rpm to about 2500 rpm in Neutral or Park and note the vacuum reading again. At 2500 rpm, the vacuum reading should be equal to or greater than the idle reading. Besides making the vehicle perform very sluggishly, an exhaust restriction will cause a substantial drop in

ALTITUDE	Inches Of Vacuum (in-Hg)
Sea Level to 1000 ft.	17-22
1000 ft. to 2000 ft.	16-21
2000 ft to 3000 ft.	15-20
3000 ft to 4000 ft.	14-19
4000 ft to 5000 ft.	13-18
5000 ft to 6000 ft.	12-17

Corrected vacuum gauge readings for higher altitudes.

USING A VACUUM GAUGE

White needle = steady needle   *Dark needle = drifting needle*

Indication: Normal engine in good condition.

Gauge reading: Steady, from 17-22-in./Hg.

Indication: Sticking valves or ignition miss.

Gauge reading: Intermittent fluctuation at idle.

Indication: Late ignition or valve timing, low compression, stuck throttle valve, leaking carburetor, throttle body or intake manifold gaskets.

Gauge reading: Low (15-20-in./Hg) but steady.

Indication: Improper carburetor adjustment or minor intake leak at carburetor or manifold. NOTE: Bad fuel injector O-rings may also cause this reading.

Gauge reading: Drifting needle.

Indication: Weak valve springs, worn valve stem guides or leaky cylinder head gasket (vibrating excessively at all speeds). NOTE: A plugged catalytic converter may also cause this reading.

Gauge reading: Needle fluctuates as engine speed increases.

Indication: Burnt valve or improper valve clearance. The needle will drop when the defective valve operates.

Gauge reading: Steady needle, but drops regularly.

Indication: Choked muffler or catalytic converter, or excessive back pressure in system. Choked muffler will exhibit a slow drop of vacuum to zero.

Gauge reading: Gradual drop in reading at idle. Reading decreases with rpm.

Indication: Worn valve guides

Gauge reading: Needle vibrates excessively at idle, but steadies as engine speed increases.

Vacuum gauge readings.

TRAINING FOR CERTIFICATION

the vacuum reading at 2500 rpm.

When the idle vacuum is low but steady, suspect air leaks first. Air leaks or vacuum leaks are common causes of rough idle, hesitation, stalling and hard starting. If you artificially enrich the mixture the right amount, an engine with air leaks should speed up and then smooth out. A propane kit with a length of hose attached, works well as a tool for finding vacuum leaks. When propane is drawn in through the leak, the engine will smooth out. Pass the hose end around the suspected areas and listen for a change in idle.

If artificially enriching the mixture makes no difference and the engine performs sluggishly, suspect a leaking EGR valve, late ignition timing or valve timing. Note that a leaking EGR valve can cause a low but steady vacuum reading. However, it can also cause a low, somewhat unsteady reading, but not as unsteady or erratic as you see with burned or sticking valves.

When in doubt, check to see if temporarily blocking off the EGR valve with gasket paper corrects the engine's rough idle, stalling and hesitation problems.

When the reading floats or slowly wanders above and below a normal idle reading, the carburetor is out of adjustment.

A vacuum reading that regularly drops to a much-lower-than-normal reading usually indicates leaking valves. When you see a substantial but very intermittent drop, suspect sticking valves.

With weak or broken valve springs, the vacuum reading usually flutters or oscillates at idle and when you raise the speed of the engine.

When the vacuum reading jumps abruptly from normal to very low, it could indicate a head gasket that has blown out between two cylinders.

CYLINDER BALANCE TEST

With a cylinder balance test, you can compare the power output of all the engine's cylinders. Some technicians call it a power contribution test because it indicates how much power each cylinder contributes to the engine. The power balance test is an important technique for solving rough idle complaints.

To perform this test, let the engine idle and short out the ignition to each cylinder, one cylinder at a time. Some technicians also short cylinders at 1500 rpm and compare the results with those of the idle test.

You can use an engine analyzer or a cylinder-shorting device to do this test safely. When in doubt, always refer to the manufacturer's recommended procedures for power balance tests. Many Ford products do this test during the Key On Engine Running test by turning off injectors.

Never short cylinders by pulling wires off the spark plugs. Open-circuiting a plug wire can give you a nasty shock and can damage the ignition system.

Because shorting cylinders dumps raw fuel into the catalytic converter, the converter could overheat. Allow about a 20-second cool-down period after you short each cylinder.

If each cylinder is producing about the same power, idle rpm will drop the same amount every time you short a cylinder. The cylinder(s) that show little or no rpm drop are either weak or dead. After you identify a weak cylinder or cylinders, you have to determine what those cylinders have in common with each other.

Two consecutive cylinders in firing order that are weak or dead often share an ignition problem. For example, the spark plug wires on these cylinders may be crossed. Or, there may be a carbon track or crack between their terminals inside the distributor cap. Sometimes, two problem cylinders share an ignition coil, as in a DIS (Distributorless Ignition System) system.

Two weak cylinders also could be related because they are the closest cylinders to an intake air leak or a leaking EGR valve. A somewhat centralized air leak such as a loose carburetor or throttle body housing can affect each cylinder to a different extent, resulting in erratic and unpredictable rpm drops during repeated power balance tests.

On a carbureted engine, watch out for rpm drops that are alternately high and low. When every other cylinder in firing order shows a high rpm drop, look for unbalanced idle mixture screws or a dirty idle circuit on one side of the carburetor. Remember that on a traditional intake system, each side of the carburetor feeds alternate cylinders in firing order.

If you have no other obvious signs of why a cylinder is weak on the power balance test, you may have to remove the valve cover and look for valvetrain wear. Before you use any measuring tools, turn the engine over very slowly and see if valve action on the weak cylinder is the same as it is on the strong cylinders. Now, check the clearances. If a rocker shaft is used, consider the possibility of shaft wear. If there are pushrods, shine a light along the length of each one. Are there any bent pushrods? A bent pushrod can cause a weak cylinder. When two cylinders right next to each other are dead, suspect a blown head gasket or a cracked head. Follow up with a cranking compression and cylinder leakage test.

COMPRESSION TEST

Cranking Compression Test

Once you have used the power

balance and cranking vacuum/cranking rhythm tests to locate a compression problem, do a compression test to determine why the cylinder is leaking compression. The two traditional ways of pinpointing a compression leak are the cranking compression test and the cylinder leakage test.

To get the most consistent and accurate results, perform a cranking compression test with the engine at normal operating temperature. Remove all of the spark plugs so the engine will crank more easily. To ensure that the engine breathes freely, remove the air cleaner and hold the throttle blade(s) wide open. Be sure the battery is strong enough to maintain the same cranking speed throughout the test. Use a battery charger if the battery is questionable.

Disable the ignition system so that the engine won't start and, where necessary, ground the coil wire to prevent it from arcing. Also, disable fuel injection systems so they don't spray fuel during the compression test. Service manuals usually list compression pressure specifications as well as allowable deviations from them.

Connect the compression gauge to the cylinder being tested, crank the engine through four compression strokes (four puffs on the compression gauge) and note how the gauge responds. Usually, cranking each cylinder through four compression strokes will give you an accurate compression reading. Pay close attention to how the gauge responds to each puff. A healthy cylinder usually builds most of its pressure on the first stroke and continues building to a good compression reading.

Wet Compression Test

During the cranking compression test, the first puff may produce weak pressure. On the second, third, and fourth puffs, the pressure may improve but never builds up to a healthy reading. When you notice this, try performing a wet compression test on that cylinder.

Squirt a spoonful of clean engine oil into the cylinder and spin the engine over to spread oil around the cylinder. Repeat the cranking compression test. If the compression improves substantially during the wet compression test, the problem may be worn compression rings, a worn piston, and/or a worn cylinder wall.

However, if the pressure is low on the first puff and remains low during a wet compression test, expect problems such as valves out of adjustment, burned valves, sticking valves, a hole in the piston, etc.

When compression is low in two adjacent cylinders, the head gasket may be blown or the block cracked between those two cylinders. Low compression on all cylinders could be a sign of worn rings on an extremely high-mileage or abused (run without enough oil) engine. It could also be an indication of valve timing that is out of specification. Compare cranking compression test results to idle vacuum readings.

CYLINDER LEAKAGE TEST

Think of the cylinder leakage test as being the last word in compression testing for a weak cylinder. In this test, you bring the piston in the weak cylinder up to TDC on the compression stroke and pump compressed air into the cylinder. Where the air leaks out shows you the location of the compression leak.

A leakage tester will compare the air leaking out of the cylinder to the amount of air you are putting into it. Generally speaking, leakage greater than 20 percent indicates a problem cylinder. If you don't use a leakage tester and are testing more than one cylinder, always use the same air pressure on each cylinder.

When the air leaks out of a cylinder, it goes to one of four places. Here's how to determine where the air is going, and why:
• Air that causes bubbles in the radiator coolant indicates a cracked head, cracked block,

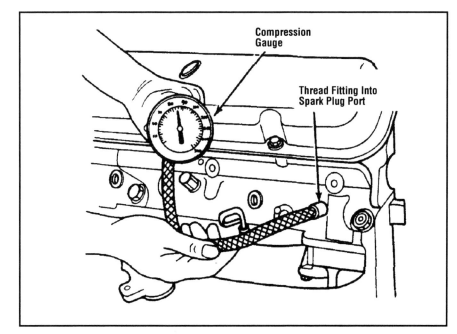

Typical engine compression tester. *(Courtesy: Ford Motor Co.)*

TRAINING FOR CERTIFICATION

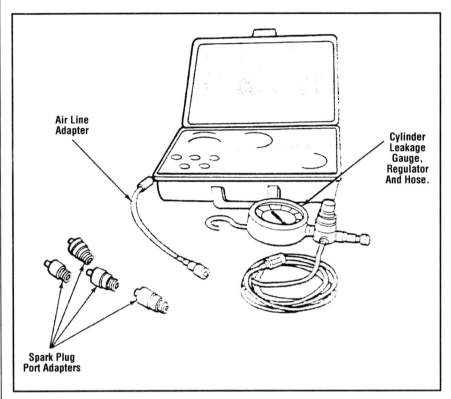

Typical cylinder leakage tester. *(Courtesy: Ford Motor Co.)*

and/or a blown head gasket.
- Air that is blowing out of the carburetor or intake system confirms that an intake valve is leaking. Be sure the air you hear in the intake isn't coming from the crankcase via the PCV valve. Plug the PCV system and listen again.
- Air that comes out of the tail pipe confirms that an exhaust valve is leaking.
- Air blowing into the crankcase indicates leaking rings and/or worn pistons. To check for this type of leak, remove the engine's oil filler cap and listen.

You can also use the leakage test to identify severe cylinder wear. After you perform the routine leakage test, do a second test with the piston at bottom dead center (BDC). A substantial difference in leakage readings from top to bottom of the cylinder indicates a badly tapered or scored cylinder.

NOTES

TRAINING FOR CERTIFICATION

CYLINDER HEAD AND VALVETRAIN DIAGNOSIS AND REPAIR

CYLINDER HEAD REMOVAL

Always wait until the engine has cooled before removing the cylinder head(s). Disconnect the negative battery cable and label and disconnect the necessary electrical connectors. Drain the cooling system.

On fuel injected engines, properly relieve the fuel system pressure. Remove the intake and exhaust manifolds from the cylinder head(s). Remove any accessories and brackets that are mounted on the cylinder head(s).

On OHC (Overhead Camshaft) engines, remove the timing belt or timing chain and sprocket(s). If the timing belt is completely removed from the engine, mark the direction of belt rotation for assembly reference. On OHV (Overhead Valve) engines, loosen or remove the rocker arms to allow pushrod removal.

Loosen and remove the cylinder head bolts in the reverse order of the torque sequence. If the cylinder head will not break loose from the engine block, first make sure all head bolts have been removed, then use a prybar inserted into a port to work the head loose.

Inspect the condition of all fasteners and components during the cylinder head removal process. Keep all parts in order so they can be reinstalled in their proper locations. Look for worn and broken parts, and damaged threads. Note the location of threads that must be repaired.

Examine the gaskets and mating surfaces of the intake and exhaust manifolds. Look for cracks, warpage and evidence of leaks and poor sealing. Carefully inspect the head gasket and look for signs of leakage.

DISASSEMBLY

The only specialty tools needed for disassembly are a valve spring compressor and a valve stem height gauge. You will also need the applicable reference specifica-

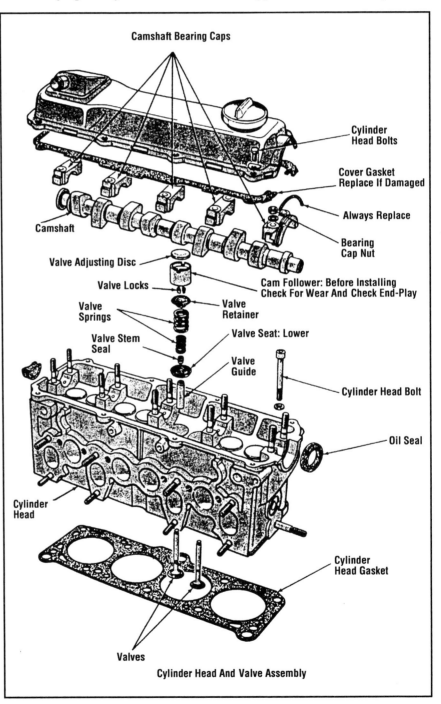

When disassembling a cylinder head, keep all parts in order so problems can be more easily diagnosed.

tions from a shop manual for checking clearances and wear.

Before you take anything apart, carefully inspect the head for obvious damage such as cracks, stripped threads, broken studs, etc. and for missing components. Also look for any markings indicating the head has been milled or fitted with a cam with oversized journals.

First remove all components necessary to gain access to the valve components as well as components that are not integral to the cylinder head. Identify the locations of all parts, prior to removal, for assembly reference. Remove all housings, covers, sensors, timing and oil pump drive components. Also, remove all fittings, studs, core plugs and oil gallery plugs. Stubborn oil gallery plugs can be heated with a torch and quenched with paraffin wax or in extreme cases, drilled and the remainder removed with a screw extractor.

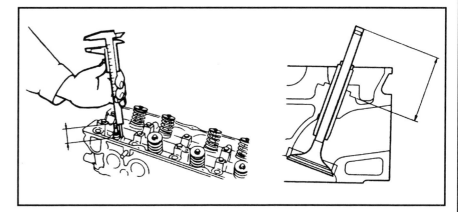
Measure the valve stem height when disassembling a cylinder head. *(Courtesy: Mazda Motor of N. America Inc.)*

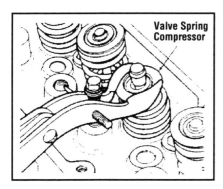
Compressing the valve spring with a valve spring compressor.

As you disassemble the valve components, line up the parts to keep them in order. This is especially important on OHC engine parts that contact the camshaft. These parts include rocker arms and bucket-type valve adjusting shims, which wear mate to the cam lobes during engine operation. When removing the camshaft from the cylinder head, make sure the camshaft bearing caps are marked for correct installation direction and position. Keeping the parts in order can also help you spot problems. For example, a rounded or mushroomed valve stem tip can also result in a badly worn valve guide or a damaged rocker arm. A bad hydraulic lifter could be the cause.

Using an appropriate valve spring compressor, compress the valve springs and remove the keepers with needlenose pliers or a magnet. Release the compressor and remove the valve spring, retainer and oil seal from each valve stem.

Before removing the valves, measure the valve stem installed height if the valve stem installed height specification is not available. This dimension will be used for later reference during assembly. Height is measured with the valve closed. It is the distance from the spring seat to the tip of the valve stem.

When a valve will not slide out but sticks in the guide, do not drive it out. A mushroomed valve stem will ruin the inside of the guide if it is pounded out. To remove a mushroomed valve without damaging the guide, hold the valve shut and file the edge of the stem tip until the valve can slide easily through the guide.

CLEANING

The selected method of cleaning should remove all grease, carbon and dirt from the head and valve components without damaging the metal. Caustic soda dissolves aluminum. When cleaning aluminum, the chemicals in a hot tank, cold tank or jet spray washer must be aluminum safe.

Thermal cleaning in an oven will bake off the grease and oil in a head leaving behind a dry powdery ash residue. This residue is removed by washing, airless shot blasting or glass beading. With aluminum heads, baking temperatures should generally be reduced from the usual 650-750°F (344-399°C) down to 400-450°F (205-232°C). Higher temperatures can cause valve seats to fall out. Also, a few heads are impregnated with resin at the factory to seal against porosity leaks. The resin will be burned out by excessive temperatures.

When cleaning heads with an airless shot blaster or glass beads, all of the blast material must be removed from the head cavities after cleaning. Glass or steel cleaning media can cause severe engine damage. A tumbler is commonly used for this purpose. Steel shot is too abrasive to use on soft aluminum heads. An alternative is to use aluminum

TRAINING FOR CERTIFICATION

shot. It's softer and is less apt to cause damage if a particle finds its way into the engine. Glass beading can be used on aluminum to remove carbon and hard, dry deposits.

Note: Glass beads are especially prone to sticking to wet surfaces inside oil galleries. This method should not be used on an aluminum head that is not completely dry.

Crushed walnut shells and plastic media are other alternative soft cleaning materials.

INSPECTION

Crack Inspection

After the head has been thoroughly cleaned, inspect it for cracks. Cracks are generally caused by thermal stress, but sometimes result from casting imperfections. The most common places where cracks form are between the valve seats, in the vicinity of the spark plug hole, in the exhaust ports, near the valve guides, and under the spring seats. When cracks extend into the cooling jacket, they often leak coolant into the combustion chamber. Due to the breakdown in lubrication, ring, cylinder, and bearing damage usually results. Cracks that are not leaking coolant are still considered a potential problem because cracks tend to grow. They may begin to leak eventually.

Crack inspection can be done using fluorescent dyes and/or magnetic crack detection equipment. Magnetic crack detection can only be done on ferrous (iron and steel) parts. Dye penetrants must be used on aluminum castings since they are not magnetic. Pressure testing can also be used to check for cracks or leaks in heads.

Magnetic particle detection is a fast and easy way to find hairline cracks in cast iron heads. The magnetic field created by the tester attracts iron powder applied to the head. A secondary magnetic field is created at the location of a crack. Additional powder accumulates around this field, outlining the crack. This technique will not reveal a crack that is parallel to the magnetic field. A second magnetic check is made by turning the tester 90 degrees. This can catch any cracks you might have missed on the first try. It is difficult to find internal cracks in water jackets and ports with this technique.

Dye penetrant is a method of crack detection that can be used on aluminum. Dye is sprayed on the surface and allowed to dry. Then the excess is wiped from the surface. A developer is sprayed on to make the cracks visible. A black light can be used with some dyes for greater visibility.

Pressure testing is another crack detection method that is good at revealing hard-to-see internal leaks. After plugging all the external openings of the water jacket, the head is lightly pressurized (usually less than 30 psi) with air. A soapy water solution is sprayed on the head to check for leaks. Bubbles highlight leaks.

When a crack is found in a head, a decision must be made before proceeding further. If there are extensive cracks that would be difficult or impossible to repair, the head must be replaced with a new or used casting. If a crack appears to be repairable, the head can be salvaged using one of several repair techniques: epoxy or heat setting caulk, pinning, lock stitching or welding.

Warpage Inspection

Check the head surface for flatness with a straightedge and feeler gauge. Check across the center of the head, at each edge and diagonally. As a rule, if there is warpage more than 0.001-in. (0.025mm) per cylinder bore, or 0.004-in. (0.102mm) in any six-inch length, the head must be resurfaced. For instance, a V6 (three cylinders on each side) would not allow over 0.003-in. (0.076mm) warpage. Resurfacing is also recommended to remove small surface imperfections or scratches that can encourage gasket failure.

Always refer to the factory specifications for flatness since some engines are not as tolerant of distortion as others are. If distortion exceeds the factory limits, the head will have to be resurfaced.

Aluminum heads are much more vulnerable to warping than cast iron because aluminum has a much higher coefficient of expansion. When mated to a cast iron block, an aluminum head tends to expand in the middle as it gets hot. Under normal conditions, the clamping force of the head bolts keeps the metal from moving excessively. But if the head overheats, it usually bows up in the middle resulting in permanent warpage and/or cracking. Severely warped aluminum heads should be straightened before they are resurfaced.

On OHC heads, warpage affects the concentricity of the cam bores. If the warpage is severe enough, it can result in rapid bore wear, loss of oil pressure and possible binding or cam breakage. OHC cam bore alignment can be checked easily before disassembly by turning the cam to see that it rotates freely. If the cam does not rotate freely, check the cam bores for wear and check the camshaft runout at the center journal. If the cam bores are OK, runout is within specification, and the camshaft does not turn easily, the head can be straightened. Straightening should be done before resurfacing or cam line boring the head. This minimizes the

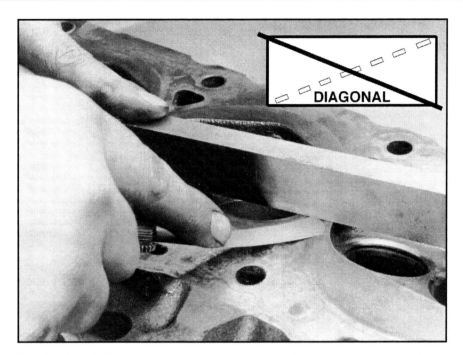

To check cylinder head warpage, place a straightedge across the head gasket surface. This should be done in diagonal directions, down the centerline of the head and on each side of the combustion chambers, checking for gaps at several points along the head.

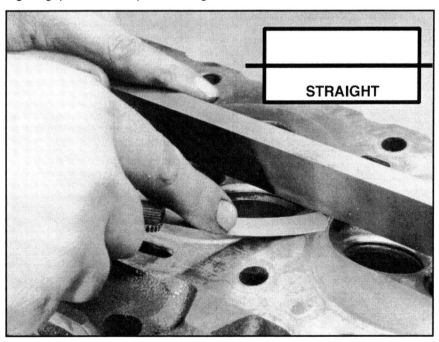

amount of metal to be removed. Line boring may be needed after straightening or welding to restore the cam bores.

Heads can be resurfaced by grinding, milling or belt sanding to eliminate warpage and restore a proper surface finish. Proper surface finish is critical for head gasket sealing. The surface finish must be correct for the type of gasket that will be used, particularly with multi-layer steel gaskets, or the gasket will not seal properly. Refer to the gasket manufacturer's specifications for surface finish requirements.

The surface finish can be checked after milling or grinding using a profilometer or comparator gauge. A comparator gauge is a metal card with sample patches of various surface textures to visually compare with the head surface.

The amount of material that can be removed from a head surface is limited. Excessive surfacing can increase compression excessively, create valve-to-piston interference problems or change the valve timing on OHC engines. To compensate for excessive metal removal, a thicker head gasket or a copper shim sandwiched together with the stock replacement head gasket can be used.

Milling the heads changes the alignment between the heads and intake manifold on V6 and V8 engines. Excessive metal removal (when the heads are milled more than about 0.030-in.) would require proportional amounts of metal to be machined off the manifold if the ports in the heads and manifold are to match up. Excess metal removal can be compensated for by installing a shim (available from gasket manufacturers) along with the head gasket.

On pushrod engines, another change brought on by resurfacing head(s) is a decrease in the distance between the lifters and rocker arms. Excessive resurfacing can upset the valvetrain geometry, causing the rocker arm to contact the valve stem tip improperly, resulting in excessive valve and guide wear. On non-adjustable valvetrain engines, excessive resurfacing can cause the location of the hydraulic lifter plungers to be too low within the lifter bodies. Shorter than stock pushrods can be used to compensate for resurfacing and restore proper valvetrain geometry.

On an OHC engine, if the cylin-

TRAINING FOR CERTIFICATION

der head or block deck has been milled, cam timing will be affected. Removal of 0.020-in. (0.508mm) from the head surface will retard cam timing by about one degree. A shim can be installed or offset cam sprockets or keyways can be used to restore correct timing.

Valves

The valves should be carefully inspected and cleaned (with a wire buffing wheel or glass bead blaster). The following are conditions which would require a valve to be replaced: cupped heads, evidence of necking (stretching and narrowing of the stem neck just above the head), pitting, burning, cracks, worn keeper grooves, too narrow a margin on the valve head, or a worn or bent stem.

Stem diameter should be measured with a micrometer and compared to specifications. Measure the valve in a worn area and compare that to an unworn area of the stem below the keeper groove. Some valve stems are ground with about 0.001-in. (0.025mm) taper. Stems on these valves are smaller at the combustion chamber end. Carefully check the keeper grooves for wear, too.

Most exhaust valves are made of a higher quality stainless steel. Always replace stainless steel exhaust valves with ones of at least the same or better steel. Check a valve to see if it is magnetic. Stainless is non-magnetic. Some exhaust valves are spin welded of

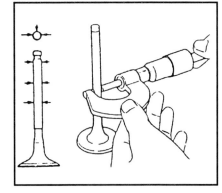

Checking valve stem wear with a micrometer.

two pieces (head and stem are different). Check the stem and valve head with a magnet. The valve head of a premium valve will be non-magnetic. The valve stems of these valves are sometimes magnetic. Some valves also have a spin welded hardened stem tip. Intake valves are occasionally of a stainless grade, too.

Valve Guides

Every head will show a certain amount of valve guide wear. Severe wear can indicate inadequate lubrication, problems with valve geometry, and/or wrong valve stem-to-guide clearance (too much or too little).

Inadequate lubrication can result from oil starvation in the upper valvetrain from low oil pressure or an obstructed oil passage. Inadequate guide lubrication can also be caused by using the wrong type of valve seal or using a positive seal in combination with an original equipment O-ring seal. Insufficient lubrication results in stem scuffing, rapid stem and guide wear, and valve sticking. Ultimately, the valve will fail due to poor seating and the resultant overheating.

Valve geometry problems include incorrect installed valve height, out-of-square springs, and misaligned rocker arms (which pushes the valve side-

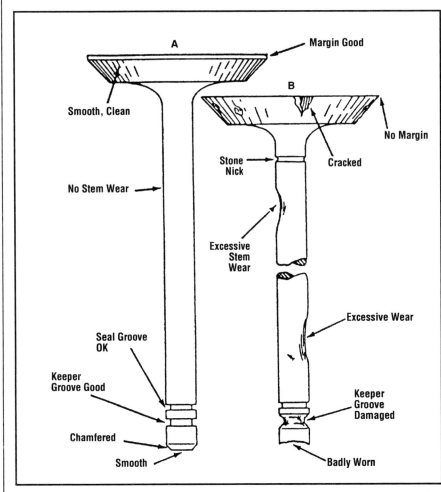

(A) An acceptable valve. (B) Problems that prevent a valve from being reused.

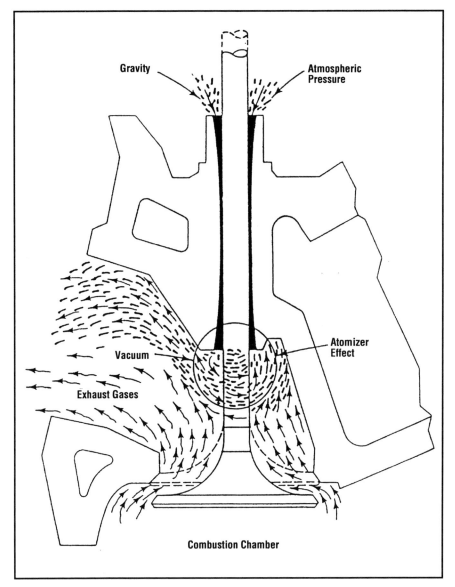

Worn valve guides and seals are a major cause of oil consumption. Guides tend to wear in a bellmouth fashion.

ways every time it opens). The resulting uneven guide wear leaves an egg-shaped hole. This leads to increased stem-to-guide clearance, poor valve seating, and premature valve failure.

A certain amount of clearance between the guide and valve stem is necessary so that oil can lubricate the stem and guide, and to allow for thermal expansion of the valve stem. Exhaust valves require more clearance than intake valves because they operate at higher temperatures and expand more. But the stem-to-guide clearance must also be tight enough to control oil consumption and to prevent exhaust gases from going up the stem and damaging the valve seal and spring. Varnish and carbon deposits can also build up on stems when there is excessive clearance. This can cause valve sticking.

Too little clearance can lead to scuffing, rapid stem and guide wear, and sticking (which prevents the valve from seating fully). A sticking valve can run hot, causing the valve to burn. The lower end of the guide is often relieved (made larger) for a distance of about 3/8-in. (9.525mm) during guide machining to prevent valve sticking.

Too much clearance can create oil control problems. Oil consumption is a problem with sloppy or worn intake guides because the guides are constantly exposed to vacuum. During the days of early automobiles, designers thought that oil could only be consumed through the intake guides. However, a lower pressure (vacuum) is created at the neck of the exhaust valve as the hot exhaust rushes past it. Oil can also be pulled down a worn exhaust guide by this difference in pressure.

Oil drawn into the engine past worn guides can foul spark plugs, cause the engine to emit higher than normal unburned hydrocarbon (HC) emissions, and contribute to a rapid buildup of carbon deposits on the backs of the intake valves and in the combustion chamber. Carbon deposits in the combustion chamber can raise compression to the point where detonation occurs when the engine is under load. Deposits on the backs of the intake valves in engines equipped with multiport fuel injection can cause hesitation and idle problems because the deposits soak up fuel, interfering with proper fuel vaporization.

Inadequate valve cooling is another problem that can result from excessive valve stem-to-guide clearance. A valve loses much of its heat through the stem. If the stem fits poorly in the guide, heat transfer will be reduced and the valve will run hot. This can contribute to valve burning, especially with exhaust valves (which don't have the benefit of intake cooling from fresh air and fuel).

TRAINING FOR CERTIFICATION

A severely worn intake guide can allow unmetered air to be drawn into the intake ports. The effect is similar to that of worn throttle shafts on a carburetor. The extra air reduces intake vacuum and leans the air/fuel mixture of the engine at idle. This can result in a lean misfire and rough idle.

Different engines have different valve stem-to-guide clearance requirements. Always refer to the factory recommendations. Typical stem-to-guide clearances range from 0.001-in. - 0.003-in. (0.025mm - 0.076mm) for intake valves and 0.002-in. - 0.004-in. (0.051mm - 0.102mm) for exhaust guides. Exhaust guides usually require 0.0005-in. - 0.001-in. (0.0127mm - 0.025mm) more clearance than the intakes because they expand more due to the hot exhaust gases. Heads with sodium-filled exhaust valves usually require an extra 0.001-in. (0.025mm) of clearance to handle the additional heat conducted up through the valve stems.

The type of guide also influences the amount of clearance needed. Bronze guides are said to be able to handle about half the clearance specified for cast iron guides or integral guides, with less tendency to seize. A knurled guide, one with oil retention grooves, or a bronze threaded liner provide better lubrication than a smooth guide. Consequently, clearances for these types of guides can be kept closer to the low side of the listed tolerance. As extra assurance against valve sticking, do not go tighter than the specified minimum valve stem-to-guide clearance.

The design of the engine's valve guide seal is also a factor to consider when determining clearances. Compared to deflector O-ring or umbrella type valve seals, positive valve seals reduce the amount of oil that reaches the valve stem. A guide with a deflector/O-ring valve seal may need somewhat tighter clearances than one with a positive valve seal to control oil consumption.

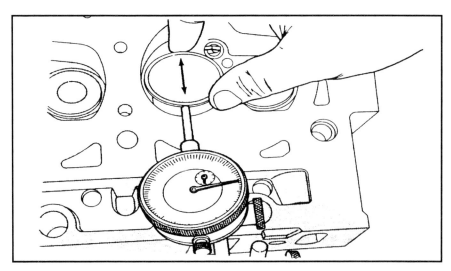
Checking valve guide wear with a dial indicator.
(Courtesy: DaimlerChrysler Corp.)

Guide clearance can be checked after cleaning the valve stem and guide with solvent and a brush to remove all gum and varnish. Insert the valve into its guide and hold it at its normal opening height. Rock it from side to side (in the same direction that the rocker arm would push on it) while checking play with a dial indicator. The amount of actual clearance is one half of the TIR (Total Indicator Reading).

When valve-to-guide clearance exceeds the specified limits, measure the valve stem with a micrometer to see if it is worn excessively. More than 0.001-in. (0.025mm) of wear calls for replacement. The inside diameter (I.D.) of the guide can be measured with a split-ball gauge and micrometer, a go/no-go gauge, or a special valve guide dial indicator. A guide will typically show the most wear at the ends and the least wear in the middle. This is called bellmouth wear.

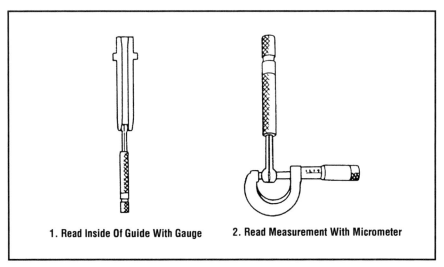
Measuring valve guide inside diameter with a split-ball gauge. The split-ball gauge is then measured with an outside micrometer.

Valve Springs

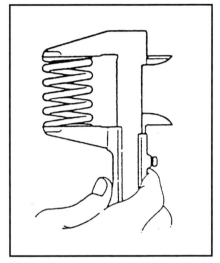

Measuring valve spring height.

Check the valve springs for tension, squareness and height. Start by checking the free-standing (relaxed) height of the springs (also called the spring free length). Individual spring height should be within 1/16-in. (1.587mm) of the original equipment manufacturer's specification. Line all of them up and position a straight edge on top of them. Any spring that is 1/16-in. (1.587mm) or more shorter than the rest should be replaced. Short springs are often found where exhaust guide clearance has been excessive. Excessive heat from the exhaust weakens the spring. Keep in mind that some cylinder heads use rotators on just the exhaust valves. If the machined spring seats on the cylinder head are all the same height, the exhaust springs will be shorter than the intakes to allow for the thickness of the rotator. In this case the free-standing height of the exhaust and intake springs should be checked separately.

Another spring check is for squareness. A bent spring will exert side pressure on the valve stem. This can cause the guide to wear abnormally or it can result in breakage of the stem tip. Place each spring on a flat surface and hold a square next to it. The spring must sit flush against the square when rotated. If the gap between the top of the spring and the square is more than 1/32-in. (0.794mm) for each inch of length, the spring is defective.

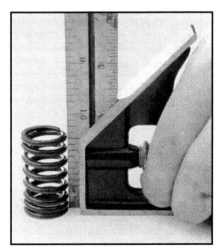

Place the valve spring on a flat surface and check it for squareness with a steel rule or square.

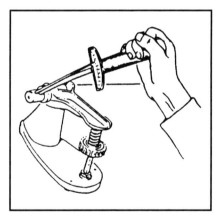

Checking valve spring pressure with a spring tester.

Check the pressure of the spring. Over time, springs lose tension from exposure to high temperatures and repeated cycling. Weak springs will allow compression loss and cause valves to run hot, shortening their life. Weak springs can also allow valves to float (remain open) at higher rpm, limiting engine speed and risking valve-to-piston contact.

Use a spring tester to check both the open and closed pressure exerted by each spring. Specifications are given for each position. At its installed height, the spring's closed pressure must be within 10% of new specifications. The same rule applies for the open pressure reading when the spring is compressed to its open height.

Do not reuse a valve spring that has shiny ends or is shiny between the coils. These are indications of a weak spring. Also do not reuse a spring that is rusted, nicked or has deep scratches. Such flaws focus stress and can cause the spring to break.

When the springs have passed inspection, carefully check the keepers, valve spring retainers and rotators and studs for wear, cracks and damage. Replace any parts found to be in less than perfect condition.

REPAIR

Valves

The valve face and valve stem tip should be reground to a new finish on a valve refacing machine. Many manufacturers specify that the valve be ground to an interference angle of 1/2 to one degree. For example, if the seat is ground to a 45 degree angle, the valve face would be ground to 44 or 44.5 degrees. Use of an interference angle improves valve seating and helps prevent carbon build-up. Always refer to the manufacturer's specifications for the proper valve face angle and dress the grinding wheel on the valve refacing machine before grinding the valve.

Clamp the valve in the refacing machine chuck and position the valve face just in front of the grinding wheel. Adjust the chuck feed so that when the valve face is

TRAINING FOR CERTIFICATION

Grinding a valve on a valve refacing machine.

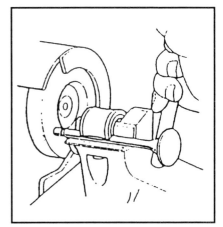

Grinding the valve stem tip.

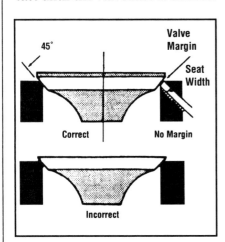

After machining the valve face, a margin of at least $1/32$-in. (0.794mm) should remain on most valves.

moved back-and-forth, the grinding wheel will clear the valve stem. Start the machine and observe the valve head. If the valve head appears to wobble as the valve rotates, the valve is bent and must be replaced.

Move the valve back-and-forth while gradually feeding the valve towards the grinding wheel, until light contact is made. Grind the valve while maintaining light contact until the valve face is smooth.

To avoid overheating the valve, make sure the coolant nozzle is directed at the valve while grinding. Grind only what is necessary. Grinding more off of the valve face leaves a narrower margin on the top of the valve head. The minimum limit of margin thickness for exhaust valves is usually $1/32$-in. (0.794mm) above the valve face. A sharp edge can cause detonation and valve burning. Some OE (Original Equipment) larger intake valves have a narrower margin. Refer to manufacturer's specifications or compare an old valve to a new one when in doubt.

The tip of the valve stem is ground to restore the surface and also to correct for minor changes in installed height that occur when the seat and valve face are reground. Valve stem height will be checked during assembly. Some manufacturers recommend that the valve stem tips not be ground at all, so be sure to refer to the specific recommendations for the engine in question.

The valve stem tip is ground by clamping the valve in the V-block of the refacing machine and passing it across the grinding wheel. When grinding a valve stem, a $1/32$-in. (0.794mm) chamfer is ground around the outside of the tip of the stem. To avoid overheating the stem, direct coolant at the stem tip while grinding.

Warning: Do not use a water-based cooling lubricant when grinding sodium filled valve stems. If the valve is cracked, the sodium inside can react explosively if it comes into contact with water!

Valve Guides

Options for repairing worn valve guides include knurling, replacement, installing guide liners, and reaming to oversize and using valves with oversized stems.

Knurling

Checking the amount of wear in the guides is the most important step to determine whether or not they can be knurled. If the guide is worn excessively, knurling may not restore the guide. Knurling decreases the guide I.D. more in the center (unworn area) than at its ends.

The knurling process consists of running a special wheel tool or ridged arbor through the guide. This leaves behind a spiral groove with raised metal on each side. The raised metal reduces the I.D. of the guide. Next, the guide is finished by reaming to its finished size.

Knurling is self-aligning, so the centering of the guide with respect to the valve seat is not lost. Knurling also allows the original valves to be reused. The other primary benefit of knurling is improved oil control, compared to a smooth guide bore. The stem-to-guide clearance for a knurled guide can be as close as 0.0007-in. (0.0178mm). When guides are knurled and clearances are tight, however, positive valve seals are not recommended.

Replacement

Replacing guides is done in both aluminum and cast iron heads. In cast iron heads, guides are often integral and a hole must be bored in the old guide to accept a pressed-in replacement guide.

To replace valve guides in aluminum heads, the heads must be heated to facilitate removal and installation without damaging the head. The guides are then usually driven out with a suitable driver and air hammer. The replacement guides are chilled and lubricated prior to installation. Most wet guides are tapered and require sealer to prevent leaks.

Replacement guides are available in a variety of materials besides cast iron. Bronze guides are generally more expensive than cast iron but usually run cooler and provide superior durability. Bronze alloys include phosphor bronze, silicon-aluminum bronze and manganese bronze.

Guide Liners

Thin tubes called thinwall valve guide liners are often installed to restore worn guides. This repair technique provides the benefits of a bronze guide surface (better lubricity, longer wear, and the ability to handle tighter clearances). Liners can be installed in heads with either integral or replaceable cast iron guides. They can be used with either standard or undersized valve stems (stems ground smaller during the rebuilding process). The process is portable, fast, and sometimes less expensive than installing new guides.

Reaming Oversize

When valves are replaced, less machining time is required. If valves are replaced with oversized stems, all that is necessary is to ream the guide to the proper oversize. Stem-to-guide clearance should be in accordance with the engine manufacturer's recommendations.

Valve Seats

The guides must be reconditioned or replaced before cutting, grinding or replacing valve seats. All seat machining is done with a pilot that centers in the valve guide. If guide work is done after seat work, the guide and seat will no longer be aligned to each other.

Valve seats should be replaced if they are badly worn or ground too deep in the head. Replaceable inserts (typically used in aluminum heads) should also be replaced if they are cracked or loose in the head. Heads with integral seats can be machined to accept a replaceable insert.

Seats can be ground with a wet or dry stone or cut with a carbide cutter. There are advantages to each method. Grinding requires at least three steps for each seat and in the case of dry stones, produces abrasive dust. Stones must

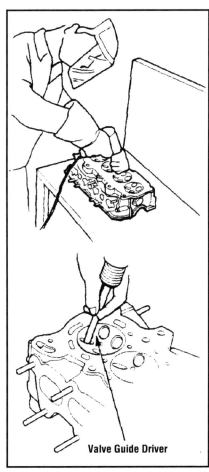

Removing valve guides from an aluminum cylinder head using an air hammer and valve guide driver. *(Courtesy: Honda Motor Co.)*

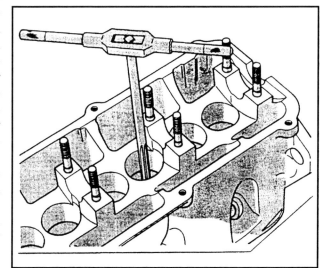

Reaming a valve guide to accept a new replacement valve with an oversize stem.

be properly dressed to maintain accuracy. Cutting is cleaner and faster, especially when all three angles are cut at once. Cutters are more expensive, however.

To grind a seat, select a wheel about 1/8-in. larger than the seat. The angle of the stone depends on the manufacturer's specifications. A typical seat face angle is 45 degrees, but occasionally 30 degrees is used (most often for intake valves). The grit of the stone depends upon the material the seat is made from. For best results and fast stock removal, use coarse stones on hard seats and a finer general purpose stone for cast iron seats.

Thread the stone onto the stone holder. Some stone holders have a ball bearing assembly inside and may be used with a lifting spring. Another type of stone holder has a brass bushing inside. A lifting spring cannot be used with this type.

Adjust the angle on the stone dressing stand to the correct setting. Dress the stone using a diamond nib, taking a small amount of material off of the stone. Attempting to remove too much from the stone at once will lift the stone and result in an uneven grinding surface. The diamond nib can also be undercut and removed from the dressing tool. Take quick strokes across the stone. Slow strokes can glaze the grinding stone (especially with finer grit stones).

Select the correct size pilot to insert into the valve guide. Slide the stone and stone holder (and lifting spring, if used) over the pilot. Grind the seat with a gentle circular motion. Be sure to be gentle and hold the stone holder in alignment to the pilot. Pushing the stone to the side can cause seat runout. If there is no lifting spring, lift the stone holder away from contact with the seat before it stops turning.

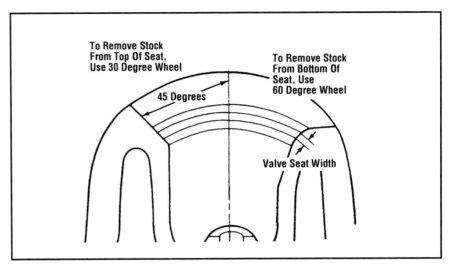

Angles used to locate and narrow a valve seat.
(Courtesy: Ford Motor Co.)

This will give a smoother finish.

Grind the seat to remove discoloration and pitting. Remove only as much material as required to clean the entire seat and remove all pits in the surface. Dress the wheel frequently (when grinding the face angle in particular). Then switch wheels and use a 30 degree stone (15 degree stone if the face angle is 30 degrees) to cut the top angle on the seat. This will locate it with respect to the valve face. The top of the valve to seat contact area should be no closer to the margin of the valve than 1/64-in. (0.397mm). If the seat is too wide, there will be no overhang. Grind as needed with the 30 degree or 15 degree stone until the correct overhang is obtained.

If the seat is too wide after grinding, narrow it by cutting the throat angle using a 60 degree stone (45 degree stone if the face angle is 30 degrees). Be careful when cutting this angle; removal of too much material from the bottom edge of the seat will call for installation of a new valve seat to correct the problem.

Dykem blue is a dye that some technicians paint on the valve seat. The valve is inserted into the guide, lightly seated, and rotated about 1/8-in. (3.175mm). A continuous blue line should appear all the way around the valve face if the valve and seat are mating properly. Open patches or breaks in the line indicate that the seat is not concentric and the low spots are not making contact.

Some technicians use another type of coloring agent when machining heads. Prussian blue is a

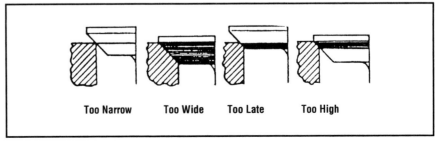

Seat width and where the valve contacts the valve seat are critical for proper sealing and long valve life.

paste that is spread on the valve face. The valve is inserted into the guide and its face is pushed into the blue paste to make an imprint. This gives an idea of the height of the valve seat on the valve face.

The procedure for refinishing a valve seat by cutting is essentially the same as for grinding. Carbide cutters are used instead of grinding stones. A seat cutting process popular in high volume shops cuts all three angles in a one-step operation. Check concentricity (seat runout) to make sure the seats were cut properly.

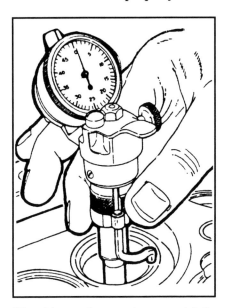

Checking valve seat concentricity using a dial indicator.

Some technicians lap valves after grinding or cutting the seats. An abrasive paste, called lapping compound, is applied to the valve or seat. A rubber suction cup on a wooden or plastic handle is attached to the head of the valve. The valve is worked back-and-forth against the seat. This process is reassuring in that a very clear pattern becomes apparent on the valve face, showing the valve-to-seat mating area. The fine lapping compound helps the seat and valve to mate exactly with each other, but the valve will expand about 0.015-in. - 0.020-in. (0.381mm - 0.508mm) when hot. This means that the lapped area will no longer match between the valve and the valve seat once the valve is hot. Lapping valve seats was a popular process in the past but it is not necessary if seat refinishing is done correctly.

ASSEMBLY

Once all of the valves, seats, and guides have been reconditioned, wash the head, valves, springs, retainers and keepers thoroughly in clean solvent to prepare for reassembly.

Valve Stem Installed Height

Install the valves in the valve guides and measure the installed stem height from the spring seat to the tip of the valve with the valve fully seated. Refer to the manufacturer's valve stem installed height specifications or, if unavailable, to the dimensions recorded during disassembly. Valve stem height that is not within specifications can change rocker arm geometry and can cause the location of the hydraulic lifter plunger to be too low within the lifter body.

If valve stem installed height is greater than specification, the valve stem tip can be ground to compensate, providing the variance is not too great. However, some valve stem tips are case hardened and should not have more than about 0.010-in. - 0.020-in. (0.254mm - 0.508mm) removed from their surface, and some manufacturers recommend that the tips not be ground at all. Grinding too much from the valve stem tip can also result in the rocker arm or cam follower contacting the keepers or valve spring retainer.

If the proper installed height cannot be obtained by grinding the tip of the stem, the valve is too deep in the head. A small amount of correction can be made with a new valve, which has more metal on its face, enabling it to sit shallower in the head. If a new valve does not bring the valve installed height within specification, a new seat will need to be installed.

Installed Spring Height

Check the installed height of the valve springs. To check the height, lightly lubricate a valve stem, slide the valve into place, and mount the retainer on the stem. Install the keepers without the valve spring for this test. Measure the distance from the machined spring seat on the head to the underside of the retainer and compare it to specifications. Both grinding the valve and grinding the valve seat result in an increase in this dimension. A valve spring insert (shim)

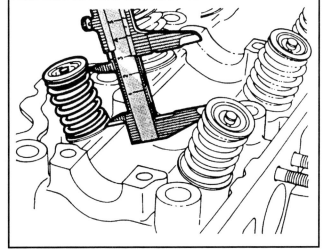

If valve spring installed height is more than specification, the spring must be shimmed to maintain adequate pressure.

TRAINING FOR CERTIFICATION

can be installed under the spring to restore the original installed height for proper spring tension.

A thin shim is found on aluminum heads to protect the softer head surface from damage. There are several thicknesses of valve spring inserts (shims) available. Common sizes for correcting spring height are 0.030-in. (0.762mm) and 0.060-in. (1.524mm). The serrated side of the insert should face the head. These serrations help insulate the spring from the hot cylinder head for longer spring life.

Valve Seals

Valve seals control oil, which lubricates the valve stems, from seeping into the guides in excessive amounts. There are three types of seals. Positive seals fit tightly around the top of the guide and scrape oil off the valve as it moves up and down. If the design of the head is such that the guides tend to be flooded with oil (and oil consumption would result), positive type seals may be specified. Positive seals are installed at the factory on almost all overhead cam engines and are installed on other heads in the aftermarket.

O-ring seals are another type of seal. They fit into the valve stem groove under the keepers and keep oil from traveling down the valve stem. These seals are used with a metal oil deflector that covers the top coils of the valve spring.

Umbrella seals (also called splash or deflector seals), ride on the valve stems to deflect oil away from the guides. These fit tightly on the valve stem and move up and down with the valve.

Valve guide seals are of several qualities. Appearance is not an indicator of quality. The least expensive seal is effective to only about 250°F (121°C). The best seals are good up to about 440°F (227°C). Valve guide seals are often included in a gasket set. Intake and exhaust seals on the same head are sometimes different. Sometimes seals are color coded. Other times they are of different shapes. Follow manufacturer's directions as to the type of seal to use on the intake and exhaust valves. A 250°F (121°C) seal used on an exhaust valve would be an unfortunate error.

When positive valve seals are installed on a head that was not designed for them, the top of the guide boss area must be machined with a special cutter. This is a common modification on high performance engines. On these applications, the valve lift should be calculated before machining, to make sure enough material is removed from the guide boss to prevent the bottom of the valve spring retainer from contacting the valve seal at maximum lift.

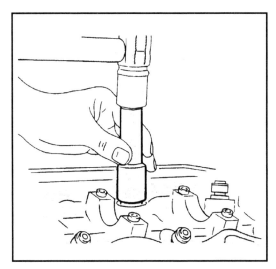

Installing positive valve stem seals. *(Courtesy: DaimlerChrysler Corp.)*

When installing positive valve seals, apply a light coat of oil on the valve stem and guide boss. Use a protective cap or sleeve over the valve stem to protect the seal as it is slid onto the valve. A special installation tool is then usually placed over the seal and the seal is tapped into place on the guide boss using a small mallet. Be sure the positive seal seats squarely on the guide boss.

With umbrella or deflector type seals, just push the seal down on the valve stem until it touches the valve guide boss. It will be positioned correctly the first time the valve opens. O-ring seals must be installed after the spring is compressed. The O-ring typically fits into an extra groove in the valve stem that is just below the keeper grooves. Lubricate it and install it squarely in its groove. Then install the keepers and release the valve spring compressor.

Final Assembly

Before installing valves in the guides, lubricate the valve stems with motor oil or assembly lube. Lubricate the valve springs, which is especially important on heads with double springs or inner harmonic dampers. Friction between the inner and outer springs can overheat and weaken the springs when the engine is first started.

After sliding each valve into place, install a shim if needed. Install the valve guide seal and put the spring and retainer into position. Then compress the spring with a spring compressor just enough to install the keepers. Be careful not to compress the spring too far. This can damage the new valve seal. Grease can be used to hold the keepers in place while releasing the valve spring compressor. Slowly release the spring compressor so there is no chance of misaligning the valve stem seals or disturbing the position of the keepers.

After all valves have been installed, tap the valve stem tips squarely with a soft mallet to make sure the keepers are properly seated.

CAMSHAFT AND VALVETRAIN

Camshaft

Clean the camshaft using solvent, and clean out all oil holes. Visually inspect cam lobes, bearing journals and accessory drive gear and fuel pump eccentric (if equipped) for excessive wear, scratches, rounded lobes, edge wear, pitting, scoring or galling. If there is any obvious wear or damage, the camshaft must be replaced.

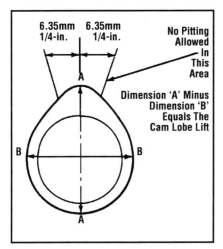

Measuring camshaft lobe wear. The difference between measurements 'A' and 'B' represents lobe lift. (Courtesy: Ford Motor Co.)

Measure each camshaft lobe height and base circle diameter using a micrometer. Subtract the base circle diameter from the lobe height to get the lobe lift. Compare to manufacturer's specification and replace the cam if necessary.

Measure the camshaft journals for out-of-round and taper and compare to specifications. If lobes and journals appear OK, place the front and rear journals in V-blocks and rest a dial indicator on the center journal. Rotate the camshaft to check straightness. If runout exceeds manufacturer's specifications, replace the camshaft.

On OHC engines, lubricate the lobes and journals with assembly lube and install the camshaft in the cylinder head. Install the bearing caps, making sure they are in their proper locations. Torque the caps to specification, in the proper sequence. If equipped, install the cam followers in their original locations.

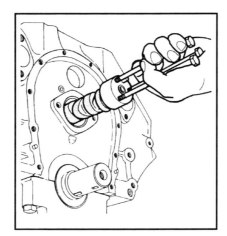

Extreme care should be used when installing the camshaft to avoid damaging the camshaft bearings.

On pushrod engines, lubricate the lobes and journals with assembly lube and install the camshaft in the block, being careful not to damage the camshaft bearings. Long bolts can be threaded into the camshaft sprocket mounting holes to provide leverage and facilitate installation. Install the camshaft thrust plate, if equipped and torque the bolts to specification.

Valve Lifters/Lash Adjusters

Remove any gum and varnish from the lifters using solvent. Inspect the face of the lifter for pitting, excessive wear or concave appearance. The face of the lifter should be convex and smooth. If any unusual wear is apparent, inspect the corresponding camshaft lobe. Inspect the lifter body for scuffing and wear and if any are found, inspect the corresponding lifter bore.

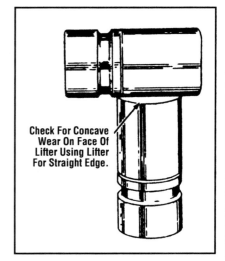

Lifters in good condition should have a convex (curved) surface on the bottom. If the bottom of the lifter is concave, it needs to be replaced.

If their appearance passes inspection, hydraulic lifters and lash adjusters can be disassembled and cleaned. Keep all components for each lifter in order; components should not be mixed between lifters. Replace the entire lifter if any components show signs of pitting, scoring or excessive wear. Replace the entire lifter if the plunger is not free in the lifter body. The plunger should drop to the bottom of the body by its own weight when assembled dry.

After reassembly, the lifters should be tested with a leakdown tester, which checks the ability of the lifter to hold hydraulic pressure. The lifter is submerged in a specified test fluid (motor oil or other fluids should not be used) and purged of air. A weighted arm, which is connected to a pointer and scale, is then placed on the lifter plunger. The time it takes for the plunger to move a certain distance (the leakdown rate) is measured and compared to manufacturer's specifications. Any lifter that has a leakdown rate that is not within specifications should be replaced.

TRAINING FOR CERTIFICATION

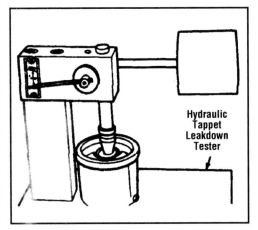

Measuring the leakdown rate of a hydraulic lifter using a leakdown tester.
(Courtesy: Ford Motor Co.)

bores in the block or head. Any nicks or scratches can be removed with light honing, if necessary. Fill hydraulic lifters with oil and purge them of trapped air, by submerging them in clean motor oil and depressing the plunger several times with a pushrod or other suitable tool. Coat the lifter face with assembly lube, lubricate the lifter body with oil and install it in the lifter bore.

Valvetrain

Inspect camshaft followers, rocker arms and rocker arm pivots or shafts for wear or damage. On shaft-mounted rocker arm assemblies, disassemble the rocker arms, springs, spacers, shafts, supports, etc. and thoroughly clean all parts. Keep all parts in order so they can be reassembled in their original locations. Measure the diameter of the rocker shafts where the rocker arms pivot and measure the inside diameter of the rocker arms. Compare measurements with specifications and replace any components that are not within tolerance.

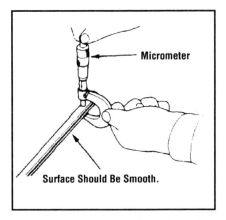

Measuring rocker arm shaft diameter.
(Courtesy: Honda Motor Co.)

Inspect the pushrod ends for wear or damage. Roll each pushrod over a flat piece of glass to check for bending. Hollow pushrods that are used for oil delivery must be clean inside.

CYLINDER HEAD INSTALLATION

Before installing the cylinder head(s), Make sure the cylinder head and block gasket mating surfaces are clean. The threads on the head bolts and those in the block must also be clean. Use a thread chaser to chase threads in the block and clean the entire head bolt using a wire wheel. Do not reuse a head bolt that is nicked, eroded or rusted.

Note: Some engines use torque-to-yield head bolts. These bolts are purposely overstretched when tightened and are usually not reused. New head bolts may be provided with the gasket set. Consult the service manual to see if torque-to-yield bolts are used.

It is recommended that if any lifters require replacement, that the camshaft and all lifters be replaced. The lifter face is convex, the cam lobe is slightly tapered (about 0.0005 - 0.0007-in.) and the lifter bore is offset to make the lifter rotate during operation. This rotation is essential because it spreads wear over a greater surface area. If the lifter fails to spin, it will destroy itself and its cam lobe. If new lifters are installed on a cam where the lobes have lost their taper, the lifters will not rotate, the load will be concentrated in one spot and the ensuing scuffing will ruin the cam and the new lifters.

Many engines are now equipped with roller lifters. Roller lifters must be checked for roller bushing wear and alignment device problems. The bushing shouldn't allow the roller excessive play on the axle pin of the lifter, and the alignment devices should retain the lifter in the bore so that the roller axle centerline stays parallel with the cam lobe centerline. This locking arrangement serves to prevent the lifter from rotating and causing the roller to slide on the lobe of the cam rather than rolling.

Clean the lifter/lash adjuster

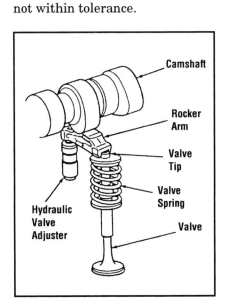

On OHC engines, be sure to inspect the cam follower and camshaft for signs of wear.

Place the head gasket on the block in the proper direction. Head gaskets are usually marked to indicate which side faces up. Older gaskets used sealer but most modern ones do not.

When installing an OHC cylinder head, be sure that the cam

36

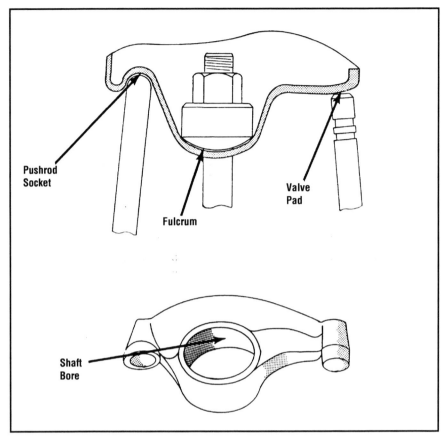

On pushrod engines, inspect wear points on the rocker arms. (Courtesy: Federal Mogul)

has been rotated until its timing mark is correctly positioned. The number one piston must be positioned at TDC so that the head can be installed. If the cam is not correctly positioned on an interference engine, a valve can be bent when the head bolts are tightened.

The tightening procedure for torque-to-yield bolts, called torque-turn, is different than the procedure for normal head bolts. A torque-to-yield bolt is tightened to a specified torque and then turned an additional amount (like 1/4 turn or 90 degrees). Torquing a bolt into yield purposely overstretches the fastener. This achieves a higher clamping force that is more uniform for each of the fasteners.

Some engine manufacturers allow torque-turn head bolts to be reused, while others do not. Follow the manufacturer's guidelines carefully on such applications.

Most head bolt specifications are for clean, lightly lubricated threads. Lubricating the threads with other than light weight oil changes the clamping force applied by the bolts at a given torque value. Torque must be reduced for proper loading if an anti-seize or other thread lubricant is used. A shop towel moistened with engine oil can be used to lightly lubricate the threads.

Hardened steel washers are placed against aluminum head surfaces. These are also used on some cast iron applications. Sealer is applied to the threads of any fasteners that extend into the cooling jackets.

Tighten head bolts in the sequence specified by the manufacturer. This allows for slight warpage of the head and stresses the block according to its engineering design. The bolts in the center of the head are tightened first. The bolts should be tightened to their full torque value in several incremental steps, rather than all at once. Leave about 10 foot pounds for the final step. For instance, if the specification is 70 ft. lbs. (94.9 Nm), tighten first to 30 ft. lbs. (40.7 Nm), then to 60 ft. lbs. (81.4 Nm), and finally to 70.

If equipped with a timing belt, inspect the condition of the sprockets, tensioner and idler pulleys. Replace any worn components. Install the camshaft sprocket and torque the bolt to specification. Align the sprocket timing marks and install a new timing belt. Adjust the tensioner according to the manufacturer's specifications and rotate the engine several times to ensure the timing marks are prop-

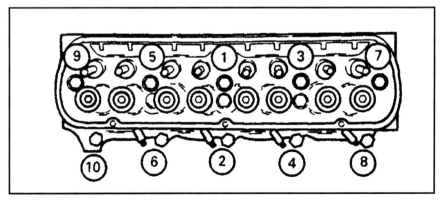

Follow the manufacturer's recommended sequence for tightening head bolts. (Courtesy: Ford Motor Co.)

erly aligned.

If equipped with a timing chain, inspect the chain and sprockets for wear and inspect the chain for stretch. If either is worn, the chain and sprockets are replaced as a set. Inspect the condition of any chain guides and tensioners and replace as necessary. Lubricate the timing chain, align the timing marks and install the camshaft sprocket and timing chain. Torque the camshaft sprocket bolt(s) to specification.

Lubricate the pushrod ends and install the pushrods through the guides into the lifters. Lubricate the rocker arm ends and pivot surfaces. If equipped with rocker arm shafts, tighten the shaft bolts gradually and evenly to the proper torque, being careful to align the pushrod ends with the rocker arms as the shaft is drawn down. If equipped with individual rocker arm pivots, align the pushrod with the rocker arm as the nut or bolt is tightened. Torque the bolt or nut or follow the valve adjustment procedure, as required.

Valve Adjustment

Some engines have a valve lash (clearance) setting that must be set within the range specified by the manufacturer. This compensates for thermal expansion. If the clearance is too tight, the valves will be held open causing compression loss and/or valve burning. If the clearance is too loose, the valvetrain will be noisy and the cam will experience shock loads and premature wear. Engine performance can also be adversely affected by the decreased valve lift and changed duration that results.

To adjust valve lash for one cylinder's cam lobes, rotate the crankshaft so that its piston is at TDC on its compression stroke. This will position the intake and exhaust valve lifters or OHC cam followers on the base circle of their respective cam lobes. With the engine in this position, adjust the clearance with a feeler gauge of the proper thickness.

Adjusting valve lash on pushrod engines calls for inserting a feeler gauge between the tip of the valve stem and the rocker arm. An adjustment nut or screw is turned to change the clearance. Occasionally there is a lock nut.

Some OHC engines use lash pad adjusters. These are shims housed in buckets that ride directly on the camshaft. They are replaced to change valve clearance. Other OHC engines use rocker arms that have an adjustment feature on one of their outer ends.

Pay attention to whether the valve lash setting recommended by the manufacturer is a hot or cold specification. With a cold specification, the lash setting requires no further adjustment once the engine is warm. A hot setting will have to be readjusted by the installer once the engine is at normal operating temperature.

Hydraulic lifter lash is adjusted by loosening the rocker arm adjusting nut until zero lash is reached. This is when the lifter plunger is extended all of the way in its travel and there is no clearance between the rocker arm and valve stem tip. Loosening the adjustment further would result in valve clearance. From the zero lash point, the rocker arm adjustment is given an additional number of turns ($3/4$ to $1 1/2$ turns is typical). The intent is to position the lifter plunger midway in its travel in the lifter body.

Changes in valve stem height resulting from valve and seat grinding or excessive cylinder head or block deck resurfacing can cause a pushrod to move the plunger lower in the lifter body. Too much increase in stem height causes the lifter plunger to permanently bottom out in its

Use a feeler gauge to measure valve clearance (lash). The clearance measuring locations for different valvetrain designs are shown here. If the lash specifications are hot, the valves will have to be readjusted after the engine has reached operating temperature.
(Courtesy: Toyota Motor Sales, USA, Inc.)

Here a technician is turning the pushrod while tightening the rocker arm adjusting nut, to feel for zero lash on a hydraulic lifter. Then the adjusting nut will be turned additionally to position the plunger in the lifter body.

bore. Also, measure the size of the cam lobe on a replacement camshaft before installation, to be sure that it is nearly the same size as the original cam lobe.

NOTES

NOTES

TRAINING FOR CERTIFICATION

ENGINE BLOCK DIAGNOSIS AND REPAIR

ENGINE DISASSEMBLY

While disassembling the engine, note the condition of the various parts. This will be helpful for diagnostic purposes and in deciding which parts require service or replacement. Wear patterns and component failures can indicate the cause of a problem and the steps necessary to prevent it from happening again.

Even when a complete rebuild is performed and most of the major components are to be replaced, it will still be necessary to inspect the old parts as they are removed. For example, an uneven wear pattern on the main bearings can be a result of misalignment in the main crankshaft bores. Machining of the main bearing bores will be necessary to restore proper alignment.

The basic procedure for engine disassembly is as follows:

1). Drain the oil from the crankcase, noting the presence of any metallic particles that would indicate a major problem. Remove and discard the oil filter.

2). Carefully remove all electrical components (sensors, sending units, distributor, ignition coil), rubber hoses, emission control components (EGR & PCV valves), accessories (alternator, starter, air pump) and accessory brackets.

3). Remove either the carburetor, throttle body or fuel rail and injectors if these parts were not previously removed. If they can remain attached to the intake manifold and will not need to be disassembled in order to remove the manifold from the head, simply remove the manifold.

4). Remove the turbocharger or supercharger (and intercooler if so equipped). Be very careful not to damage the blades on a turbocharger impeller or turbine wheel.

5). Note areas of heavy grease deposits or oil or coolant leaks. Use degreaser or detergent to clean the outside of the engine with a pressure washer. Cleaning should be performed in a special area designed to trap toxic materials for later disposal.

6). Remove the intake and exhaust manifolds. Inspect the exhaust manifolds for cracks. Check the exhaust crossover and EGR passages for carbon deposits. Clean these passageways, if necessary.

7). Remove the valve cover(s) and the upper valvetrain components while checking for worn or broken parts. Keep all mated parts together. If they are to be reinstalled later, they will have to go in the same position. Also, if a part is worn, a cause will have to be determined and a course of action taken to prevent it from happening again. The corresponding part that caused the wear will also have to be carefully inspected.

8). On OHC engines, remove the water pump, crank pulley and damper. Remove the timing cover and remove the timing belt or chain. Timing belts should not be reused but belt sprockets are usually reused. Visually inspect the belt sprockets for unusual wear. Feel the idler roller bearing used with a timing belt. These are usually replaced with a new belt.

On an OHC engine with a chain drive, the chain guide and tensioner are coated with a layer of synthetic rubber. These parts usually experience wear and are routinely replaced. A master engine kit for these engines usually includes a new timing chain, sprockets, chain guides and a tensioner. These parts are also available packaged together as a timing set.

9). Remove the cylinder head and gasket. Inspect the head gasket and cylinder head surface carefully. The location of coolant or combustion leaks can often be determined during this inspection. Questionable areas can be scrutinized more carefully while checking warpage. Discard the gasket after inspecting it.

10). On pushrod engines, remove the water pump, crankshaft pulley and damper, and timing cover from the front of the engine. Timing chains on these engines typically stretch in use. Note how much a chain has stretched before removing it. Turn the crankshaft clockwise to take up slack on the right side of the chain. Mark a reference point on the block at the approximate midpoint of the chain. Measure from this point to the midpoint of the chain.

Rotate the crankshaft counterclockwise to take up slack on the left side of the chain. Force the left side of the chain outward with your finger and measure the distance between the reference point and the midpoint of the chain. The timing chain deflection is the difference between the two measurements. Replace the chain and sprockets if deflection exceeds specifications.

11). Inspect the timing sprockets for visible wear. When they are worn, the chain and both sprockets are replaced as a set. A new timing chain should always be installed on new sprockets.

12). Check the camshaft end-

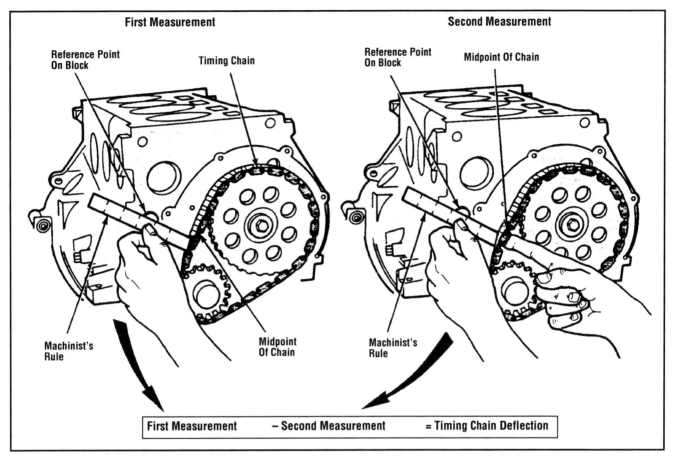

Measuring timing chain deflection. *(Courtesy: Ford Motor Co.)*

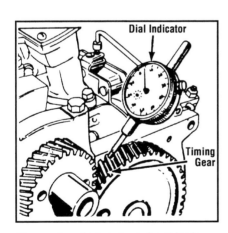

Measuring timing gear backlash.

play with a dial indicator. If endplay is excessive, the camshaft and/or thrust plate are worn. On an engine with a timing gear drive, a dial indicator or feeler gauge can be used to check gear backlash. The timing gears should be replaced if backlash exceeds specifications.

13). Check the flywheel for wear and runout. Look for heat cracks on the clutch face. Also check for excessive runout. Either of these conditions will require replacement of the flywheel. Look for missing or damaged gear teeth on the starter ring gear. The ring gear can be removed and replaced on some manual transmission flywheels. When starter teeth are damaged on an automatic transmission flexplate, the flexplate is replaced.

When a flywheel clutch surface has hot spots (blue or black marks), scoring or warpage, it can often be successfully resurfaced using a flywheel grinder. Runout is checked by mounting a dial indicator to read off of the flywheel's clutch surface and then turning the flywheel one complete revolution. If the maximum runout between the highest and lowest point exceeds

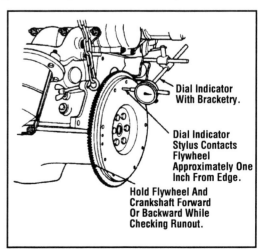

Measuring flywheel runout with a dial indicator. *(Courtesy: Ford Motor Co.)*

TRAINING FOR CERTIFICATION

the manufacturer's specified value, the crank flange or flywheel mounting surface could be bent.

Use a straightedge to check for warpage across the clutch face of the flywheel. If warpage exceeds the manufacturer's specified value, the flywheel will have to be resurfaced. Also check the pilot bearing in the rear of the crankshaft for wear.

Note: It is a good habit to always mark the flywheel and clutch parts before disassembly. On externally balanced engines, do not unbolt the flywheel from the crankshaft until you mark its index position with a punch. This is necessary to maintain proper engine balance.

14). On a pushrod engine, remove the cam and lifters. If the lifters are to be reused, they must be kept in order so that each one may be put back in the same bore from which it was removed. Pushrod engines usually experience excessive wear to the cam and lifters. Cam lobes are tapered and lifters are convex. This is so the lifters will spin and dissipate loads on them. Look for wear on the bottom edges of the cam lobes. One worn lobe is cause for replacement of all of the lifters and the camshaft. New lifters must always be used with a new cam, and vice versa.

15). Remove the oil pan and gasket and discard the gasket. The oil pump and pickup screen are removed next. Note the presence of any debris in the pickup screen, which would call for its replacement. Most screens have a built-in pressure relief valve in case they become restricted. When the engine has been run at high speed while the screen has been plugged by debris or heavy oil, the screen can become permanently deformed. This leaves the relief valve open, which can allow debris to enter the pump. Oil pump pickup screens are usually replaced. However, if the screen is to be reused, it must be thoroughly cleaned.

Check the oil pump gears for wear. A new or rebuilt oil pump is included in most master engine kits. Pumps that are enclosed within the oil pan are commonly replaced during a rebuild. Replacing a defective oil pump later is often a difficult proposition.

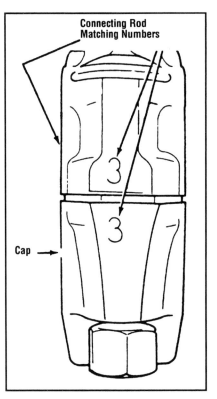

An example of cylinder identification marks stamped on a connecting rod.

16). Remove the pistons and connecting rods. Look for numbers stamped on the sides of the connecting rods indicating their cylinder number. If there are no numbers, use number stamps to mark the position of each connecting rod according to its cylinder number. Stamp the numbers on both sides of the big end parting face, so the rod caps will not be mixed up during removal.

Before the pistons are removed, check for a ridge at the top of the cylinder bore. This ridge occurs because the piston ring does not travel all the way to the top of the bore, thereby leaving an unused portion of cylinder bore above the limit of ring travel. This ridge will usually be more pronounced on high mileage engines. If the ridge is severe, it must be removed with a ridge reamer before pushing the pistons out the top of the block.

To remove the ridge, place the piston at the bottom of the bore, and cover it with a rag. Cut the ridge away using a ridge reamer, exercising extreme care to avoid cutting too deeply—that is, more deeply than is necessary to just even out the top of the bore. Also, don't allow the cutter bit to come high enough out of the bore to tilt and chamfer the top edge of the cylinder wall. Remove the cutter. Return the piston to TDC. Remove the rag, and then carefully remove any cuttings that remain on top of the piston.

Note: A severe ridge is an indication of excessive cylinder bore wear. Before removing the piston, check the cylinder bore diameter with a bore gauge and compare the measurement with specification. If the bore is excessively worn, the block will be rebored for oversize pistons so removing the ridge is not necessary.

Remove the piston and rod assemblies one at a time. Remove the rod bolts or rod bolt nuts, as required, and remove the connecting rod cap. Slip pieces of rubber hose over the rod bolts to keep the bolt threads from damaging the crankshaft during removal. Push each assembly upward and remove from the top of the bore. After the piston and connecting rod assembly is re-

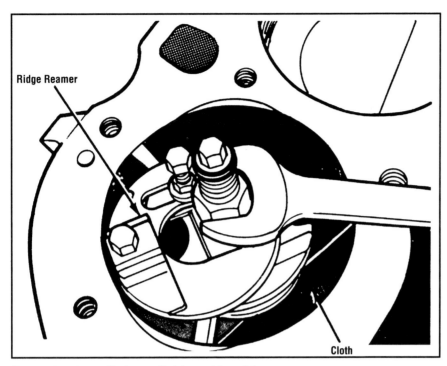

Removing the cylinder wall ridge with a ridge reamer.

moved from the block, reinstall the rod cap.

Inspect the pistons and connecting rods for cracks, damage and wear patterns that might indicate other problems. Scuffed pistons, for example, might be the result of inadequate lubrication, insufficient piston-to-cylinder clearances and/or overheating. A fuel system problem can contribute to a piston overheating condition called four corner scuffing. Damaged ring lands are often a result of a detonation problem. A hole burned in the top of the piston can also be caused by detonation.

17). Remove the crankshaft. During manufacture, each main cap is finish machined at the same time as its corresponding bore half in the block. On most engines, the cap's position in the block is stamped or cast into the main cap. If no identifying numbers can be seen, before removing any of the main caps, use number stamps to mark their positions in the block. The caps must be reinstalled in their original positions.

Remove the main bearing caps, lift out the crankshaft, and remove the bearings. Mark the back of each bearing with a felt marker for future reference.

18). Remove and inspect the camshaft bearings. Note any wear of the camshaft bores that would indicate a need for line honing or line boring. If the center bearing shows more wear than the end bearings, either the camshaft is bent or the block is warped.

19). Remove the core plugs and oil gallery plugs. Core plugs can be removed using a punch to twist the plug in its hole so it can be pulled or pried out. They can also be removed by knocking a hole through them with a chisel and prying them out. Be careful not to damage a cylinder by forcing a core plug into the coolant jacket. A cylinder can easily be distorted.

Small oil core plugs can be removed by drilling a small hole in the plug. Turn a sheet metal screw into the hole and pull the plug out with pliers or a slide hammer. Threaded pipe plugs are easiest removed by first heating them with a torch and then quenching with wax.

Remove any remaining studs or bolts.

20). Spread the main and rod bearings out in the same positions as they were previously installed in the block. Note any unusual wear patterns. Check to see if the center main bearings show more wear than the ones toward either end of the crankshaft. This could indicate that the crankshaft is bent. It will have to be straightened or replaced. If the main bearing bores of the block are misaligned due to block distortion, the upper or lower bearing halves will show wear. The centermost bearing will usually show the most wear. This will re-

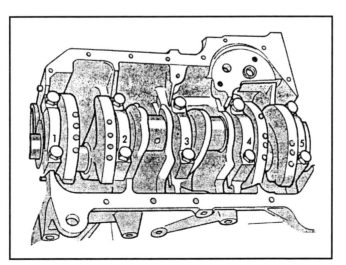

Be sure that the main bearing caps are marked before removing them. They are not interchangeable and must be returned to their original positions.

TRAINING FOR CERTIFICATION

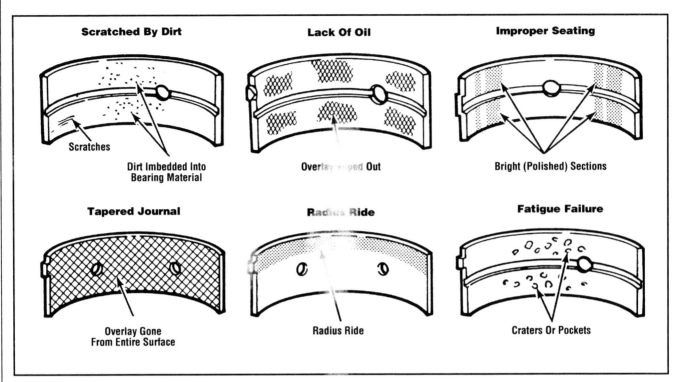

Reading bearings after the engine has been disassembled can help you diagnose engine problems that need repairs.

quire align honing or boring of the main bearing bores. If only the upper front bearing is worn, too tight of a belt adjustment is indicated.

If the engine had a knock, the rod bearing(s) farthest from the oil pump might show the most wear. This would indicate an oil supply problem.

Oil supply problems can result from:
• A loose or leaky oil pump pickup
• Restricted pickup screen
• Worn oil pump
• Excessive main bearing clearance
• Low oil level
• Too much oil in the crankcase.

If there is too much oil in the crankcase, the spinning crankshaft can churn the oil into foam. The foam is drawn into the oil pump and a lack of crucial lubrication results.

When a single rod bearing shows excessive wear but the rest are fine, several things could be the cause:
• An obstruction in the crankshaft oil passage(s), which supplies the bearing
• Out-of-round rod bore
• Stretched rod bolts
• Loose or improperly torqued rod nuts
• Twisted or bent rod
• Out-of-round or rough connecting rod journal
• Misassembly, including incorrect bearing crush fit, incorrect oil clearance, installation of a wrong-sized bearing or dirt behind the bearing back
• Detonation or preignition in the related cylinder.

Burned bearings and/or discolored crank journals are signs of inadequate lubrication. Possible causes include clogged oil passages, internal oil leaks, a weak or broken oil pump, improperly installed pickup tube, engine overheating, an oil relief valve stuck in the open position or low oil level due to leaks or neglect.

Particles embedded in the bearings or badly scored bearings are due to oil contamination. Contamination can result from infrequent oil and filter changes, oil bypassing a clogged filter, unfiltered air entering the crankcase through the PCV system or metal debris from damaged components elsewhere in the engine.

Upper rod bearings and lower main bearings which show signs of fatigue stress (flaking separation of bearing layers from the steel backing) may be a symptom of detonation or preignition.

Corroded bearings are the result of acid buildup in the crankcase due to infrequent oil changes, a restricted PCV system or poor quality oil.

CLEANING

After the engine has been disassembled, the cylinder block, crankshaft and related parts must be thoroughly cleaned. Sludge, carbon deposits, grease, dirt, scale and other debris are removed. The selected cleaning technique must also clean hidden areas of the block such as oil passages and water jackets.

The most common technique for cast iron cleaning used to be soaking the block, heads and other parts in a hot tank filled with a solution of caustic soda and hot water, or placing the parts in a jet clean cabinet where they would be cleaned by high pressure jets of caustic solution while rotating on a turntable. Due to environmental concerns, however, thermal cleaning has become a popular alternative. With this method, cast iron parts are placed in a high temperature bake oven or pyrolytic oven set at about 750°F (399°C), and the oven burns out the grease and other contaminants. The remaining ash and scale is then removed by washing or by shot blasting. After shot blasting, the part is tumbled in a tumbler to remove leftover shot.

Note: It is extremely important that no shot blast media be left in the block where it could find its way into the oil system.

Aluminum parts can be cleaned chemically; however, caustic solutions are too strong for cleaning aluminum parts. A detergent based or aluminum safe chemical must be used to chemically clean aluminum. If aluminum parts are thermal cleaned, a lower temperature setting, about 450°F (232°C), must be used. Steel shot is too abrasive to use on soft aluminum. An alternative is to use aluminum shot, crushed walnut shells or plastic media. Glass beading can also be used. If the pistons are cleaned with glass beads, only remove the carbon from the top of the pistons and do not glass bead the ring grooves. Wrap masking tape around the ring band to protect the ring grooves prior to glass beading.

Small parts are typically cleaned in cleaning solvent (cold tank) or chlorinated hydrocarbon (carburetor cleaner). Bead blasting with glass, aluminum or soft media can also be used to clean small parts and aluminum castings. As with steel shot cleaning, it is very important that cleaning media does not remain inside the engine. Glass beads are especially prone to sticking to wet surfaces inside oil galleries.

INSPECTION

Crack Inspection

After the block and other parts have been cleaned, they are inspected for cracks. Crack inspection of the crankshaft, connecting rods, main caps, block and heads can be done using fluorescent dyes and/or magnetic crack detection equipment.

Magnetic crack detection can only be done on ferrous (iron and steel) parts. Wet magnetic particle inspection is normally used to check crankshafts, camshafts and rods. Dry magnetic particle inspection is used to check blocks and heads.

Dye penetrants must be used on the pistons and aluminum castings since they are not magnetic. Pressure testing can also be used to check for cracks or leaks in the heads or block.

Cracked blocks are not necessarily replaced. A cracked cylinder can often be successfully repaired by installing a sleeve. Cracks or holes in external water jackets and deck surfaces are tougher to fix. They require welding, gluing or pinning.

Warpage Inspection

The deck surface of the cylinder block must be clean and flat if the head gasket is to seal properly. A blown head gasket or a buildup of carbon on the deck surface is an indication that the deck surface is uneven. Check for warpage with a straightedge and feeler gauge. Warpage of more than 0.001-in. (0.025mm) per cylinder bore, or 0.004-in. (0.102mm) in any six-inch length, requires that the deck be resurfaced. For instance, a V6 (three cylinders on each side) would not allow over 0.003-in. (0.076mm) warpage. Resurfacing is also recommended to remove small surface imperfections or scratches that can encourage gasket failure.

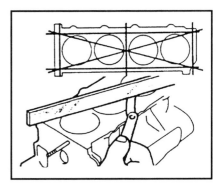

Check the deck surface of the block for warpage with a straightedge at these positions.

The deck can be resurfaced by grinding or milling to eliminate warpage and restore a proper surface finish. Proper surface finish is critical for head gasket sealing. The surface finish must be correct for the type of gasket that will be used, particularly with multi-layer steel gaskets, or the gasket will not seal properly. Refer to the gasket manufacturer's specifications for surface finish requirements.

Cylinder Bores

Cylinder bores wear in two ways, taper and out-of-round. Tapered cylinders are worn more at the top, where ring loading is the greatest. Cylinders can be out-of-round as a result of the motion of the piston rocking on its wrist pin.

Visually inspect the cylinder bores for nicks, scratches, scoring

TRAINING FOR CERTIFICATION

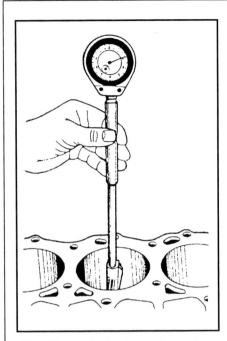

Checking cylinder bore dimensions with a dial bore gauge.

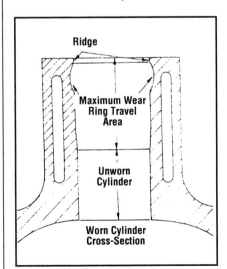

Cylinders develop the most wear at the top where ring loading is the greatest.

or scuffing. Measure the cylinder bore using a dial bore gauge. Measure for wear, taper and out-of-roundness. Refer to the manufacturer's specifications for bore size to set the gauge or zero the gauge on an unworn portion of the cylinder. Any movement from zero on the gauge represents cylinder bore wear.

Measure the cylinder bore at the lowest point of piston ring travel, then gradually move the gauge to the uppermost point of ring travel. Measure at points perpendicular to and parallel to the crankshaft (90 degrees apart). The difference in the upper and lower measurements is the cylinder bore taper; the difference in the 90 degree apart measurements is the cylinder bore out-of-roundness.

Compare your measurements with the manufacturer's specifications. If cylinder bore wear is within specification, then honing is all that will be necessary to prepare the cylinder wall surface for new rings. However, if wear is beyond specification, the cylinders will have to be bored, honed and new pistons installed. Consult the engine and/or piston manufacturers for available oversizes.

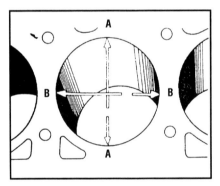

Measure the cylinder bore at both points A (perpendicular to the crankshaft) and B (parallel to the crankshaft) to determine bore wear and roundness. Measurements should be taken at both the top and bottom of the cylinder to check for taper wear.

Cylinder Sleeves

Some engines have dry cylinder sleeves or liners that are either slide-fit or pressed-fit in the block. They are called dry sleeves because coolant does not contact the outside of the sleeve. Both types can be replaced if the sleeves are worn, but require the use of special pullers or a press. Sleeves may also be used to repair bad cylinders that are cracked or worn in regular integral liner blocks.

Wet sleeves are so called because they are exposed to coolant. They have soft metal rings at the top, and rubber rings at the bottom to seal the sleeves in the block. Wet sleeves can usually be replaced without special pullers or a press.

Main Bearing And Camshaft Bearing Bores

Camshaft and crankshaft bores should be checked for alignment and distortion. Bore diameter can be checked using a bore gauge. As a rule, if the bores are more than 0.0015-in. (0.0381mm) out-of-round, they should be align bored or align honed. When cam bores are machined larger, oversized bearings are installed.

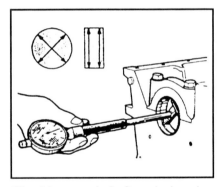

Checking crankshaft main bearing bore diameter.

Alignment can be checked using a straightedge and feeler gauge. A deviation of more than 0.0015-in. (0.0381mm) will require machining to correct.

Crankshaft

Clean the crankshaft journals thoroughly with solvent to permit a thorough inspection. Use pipe cleaning brushes to ensure that all oil holes and passages in the crankshaft are open and free of sludge. Check the journals and

Using a straightedge and feeler gauge to check main bearing bore alignment.

thrust faces for scoring, nicks, burrs or any form of roughness. If any journals are discolored (blue or black) this indicates overheating, due to a failed bearing. The metallurgy of a crankshaft can be changed when it overheats and the crankshaft will have been weakened; it will have to be replaced or heat treated as part of a repair. If the crankshaft is from a manual transmission vehicle, check the condition and fit of the transmission pilot bearing.

A crankshaft journal that looks acceptable to the naked eye might still be out-of-round, tapered or worn. Measure the crank journal diameter with a micrometer at either bottom or top dead center, and again at 90 degrees in either direction. Rod journals typically experience the most wear at TDC so comparing diameters at the two different positions should reveal any out-of-roundness. The traditional rule-of-thumb for journal variation says up to 0.001-in. (0.025mm) is acceptable, but many of today's engines cannot tolerate more than 0.0002-in. - 0.0005-in. (0.0051mm - 0.0127mm) of out-of-roundness. Be sure to check specifications.

Bearing journals can experience taper wear (one end worn more than the other), barrel wear (ends worn more than the center) or hourglass wear (center worn more than the ends). Measure the journal diameter at the center and both ends. The journal diameter itself should be within 0.001-in. (0.025mm) of its original dimensions, or within 0.001-in. (0.025mm) of standard regrind dimensions for proper oil clearances with a replacement bearing.

An easy way to tell that a journal has been previously reground is to look for machinist markings stamped on the crank. A number 10, 20, or 30 stamped near a journal tells you the journal has already been machined undersize. Of course, if there is any question, measuring will give you the actual dimension. Further regrinding may be out of the question, depending on how badly the crank is worn. Replacement bearings are available for most crankshafts machined up to 0.030-in. (0.762mm) undersize. In addition, if the crankshaft is still at standard size, worn uniformly and still useable, bearing shell sets of 0.001-in. (0.025mm), 0.002-in. (0.051mm), and 0.003-in. (0.076mm) over standard size are sometimes available to correct excessive bearing clearance.

A bent crankshaft can be diagnosed by inspecting the main bearings for wear during disassembly. Place the crankshaft in V-blocks and rotate it while watching for wobble at the center main journal with a dial indicator. If run-out exceeds limits, the crank will need to be straightened or replaced.

If the journals are free from flaws and wear, taper, out-of-roundness and runout are within specification, the crankshaft journals can be polished and the crankshaft returned to service. However, if the crankshaft fails to meet these criteria, it will have to be reground 0.010-in. (0.254mm), 0.020-in. (0.508mm) or 0.030-in. (0.762mm) undersize. Most journals will clean up at 0.010-in.

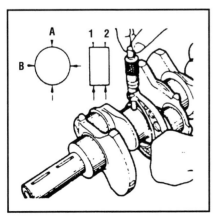
Measure the crankshaft journals with a micrometer.

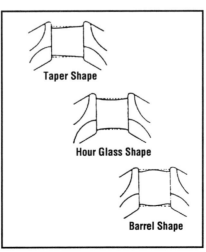
Worn crankshaft journals.

49

TRAINING FOR CERTIFICATION

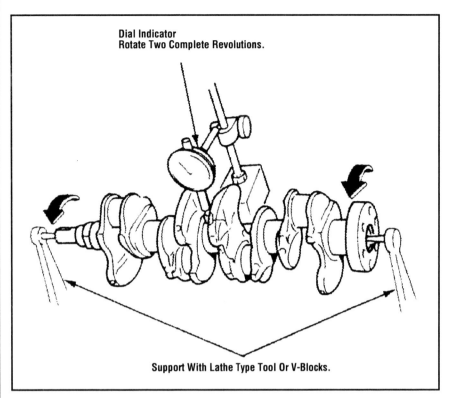

Measuring crankshaft runout with a dial indicator. *(Courtesy: Honda Motor Co.)*

(0.254mm) undersize. If it takes more than 0.030-in. (0.762mm) to restore the journals, the crank may have to be replaced or built up by welding metal onto the journals. After grinding, the machinist will chamfer the oil holes to promote good oil flow to the bearings, and then polish the journals with a fine grit emery belt.

Pistons

Clean the pistons and carefully inspect them for wear and damage. Note the condition of the rings. Piston rings can wear in normal use or because of a lack of proper lubrication or from dirty air or oil. Abrasive wear accounts for most ring failures. Overheating is also a major cause of failed rings.

If the rings are to be replaced, remove the old rings from the pistons with a ring expander. Do not spiral rings off or on. Check the

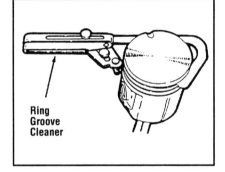

After the rings have been removed, clean and inspect the ring grooves.

ring lands for wear, cracks or damage. Clean the ring grooves with a ring groove cleaner. Be very careful not to damage the ring grooves.

Do not use the wire wheel on a grinder or attempt to clean the pistons in a hot tank filled with caustic soda. If pistons are cleaned with a glass bead blaster or in carburetor cleaner (chlorinated hydrocarbon), the connecting rods and wrist pins must be removed first. If the piston is cleaned with glass beads, only remove the carbon from the top of the piston and do not glass bead the ring grooves. Wrap masking tape around the ring band to protect the ring grooves prior to glass beading.

Note: Do not soak the piston and rod assemblies in solvent as the wrist pin can seize to the piston. Always disassemble the pistons from the pins and rods if the pistons are to be soaked in a chemical cleaner.

Use penetrating dye to check the piston for invisible hairline cracks around the wrist pin boss, ring lands, crown and skirt areas. Replace any piston found to be cracked. Note unusual wear patterns. Scoring on one piston skirt can be caused by inadequate lubrication due to prolonged idling or engine lugging. Scuffing on both piston skirts is usually an indication of insufficient piston-to-bore clearance after a rebore. A cooling system problem will often cause scuffing of the skirt on the sides near the pin due to excessive expansion. Holes or melted spots on the piston indicate a preignition or detonation problem. Cracked ring lands indicate detonation. A slanted wear pattern on the side of the piston indicates a twisted connecting rod. Wear above the wrist pin can be due to a bent connecting rod. Bent and twisted rods can be accompanied by misalignment wear of the connecting rod bearings.

Piston wear can be checked by measuring the diameter of the piston with a micrometer and comparing it to specifications. Pistons are cam ground (slightly oval shaped to allow for thermal expansion). Their largest diameter is at a point about one inch

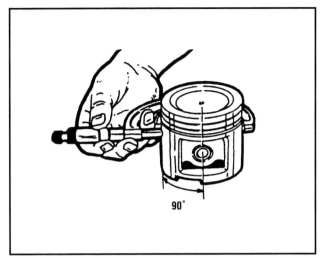
Measuring piston diameter with a micrometer.

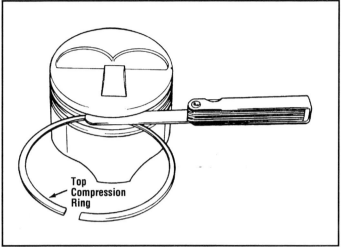
Checking ring side clearance. *(Courtesy: GM Corp.)*

below the oil ring groove, perpendicular to the wrist pin. However, check the manufacturer's specifications for specific measuring locations. If the diameter is smaller than the specified service limits, the piston must be replaced.

Check the ring grooves for wear. Every time the piston changes direction in its up and down travel, the rings are forced against the piston lands. Over time this pounds the lands out of shape. Check the side clearance of a ring by inserting a new ring backward into the land and measuring its clearance to the groove with a feeler gauge. Compare your measurement with specifications. Roll the ring around the circumference of the entire ring groove, while checking for bind.

When a groove is damaged or worn excessively, the usual practice is to replace the piston. However, if an expensive piston in otherwise good condition has excessive ring side clearance, the top groove can be cut wider using a special lathe. A spacer is then installed with the new ring.

When a piston has a worn skirt, the piston can rock back-and-forth in its bore, accelerating ring, piston and bore wear. Skirt clearance can be checked by measuring the piston diameter with a micrometer and comparing it to the diameter of the bore. Another way of checking clearance is to insert the piston part way into the bore and use a feeler gauge to check piston-to-bore clearance. The feeler gauge must be at the largest part of the skirt.

Wear in the wrist pin area can cause an engine knock as the piston rocks and twists on the ends of the wrist pin. When a pin fits loose in the piston, the piston pin bore and rod small end can be resized to fit an oversize wrist pin.

Once piston inspection has been completed, determine whether or not the pistons will be removed from the connecting rods. If the pistons are to be replaced or connecting rod service is required, the wrist pins and connecting rods must be removed. Pressed fit pins are removed in a press with suitable fixtures to properly support the piston. The pins in full floating pistons are retained with clips located in the piston pin bore. These clips are discarded after removal and should not be reused.

Connecting Rods

Connecting rods can suffer from a variety of problems. Cracked rods should be replaced. Worn piston pin bushings can be replaced and out-of-round bearing bores can be remachined. Bent or twisted rods can be straightened but are most commonly replaced with an inexpensive rebuilt rod.

Inspect the old connecting rod bearings for uneven wear that could indicate that a rod is bent or twisted. Check connecting rod straightness on a rod aligning fixture.

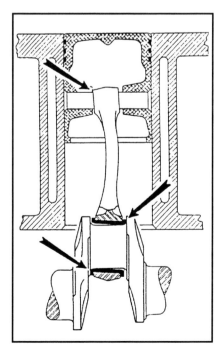
Load areas caused by a bent connecting rod.

TRAINING FOR CERTIFICATION

A rod alignment fixture is used to check for bends and twists in connecting rods.
(Courtesy: Sunnen Products Co.)

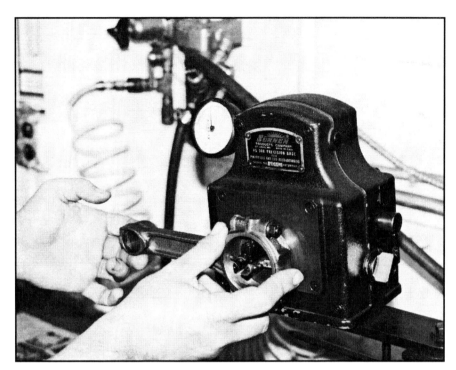

Measuring connecting rod big end bore diameter.

The connecting rod big end bore concentricity must be within 0.001-in. (0.025mm). This is checked on a gauge that reads to 0.0001-in. (0.0025mm) and compared to the manufacturer's specifications. Rods that do not meet specification are resized on a rod reconditioning machine. The rod bolts are removed and the rod and cap ground on a cap grinding machine. The minimum amount of material is removed, usually no more than 0.003-in. (0.076mm). The rod is then reassembled and the big end bore honed to size.

Pressed fit pins must have the right amount of interference in the connecting rod or they can come out and damage the cylinder wall. When the wrist pin is pressed into the rod, the pin bore is usually 0.001-in. (0.025mm) smaller than the pin. If the small end bore is not within specification, the small end bore and the corresponding piston bore can be honed and an oversize wrist pin installed.

When the piston uses full floating pins, bushings are usually installed in the small end bore. If a bushing is damaged or if the bore is not within specification, the old bushing is driven out and replaced. The new bushing is then honed to provide the proper clearance to the wrist pin.

Ideally, a new rod should be matched as closely as possible to the weight of the original to maintain engine balance. This is not as critical in a low rpm, light-duty engine as it is in a high rpm or heavy-duty engine. It is more critical in V-type engines than it is with in-line engines.

CYLINDER HONING

Honing applies a surface finish to the cylinder walls so that the rings can seat with minimal wear. This surface finish must retain enough oil to provide a bearing surface for the rings and keep the rings lubricated, but not retain so much oil that oil consumption becomes a problem.

The type of finish applied is determined by the type of hone used and the grade of abrasive. This in turn is determined by the type of ring that will be used and the amount of cylinder wall material that will be removed. A flexible hone, sometimes called a glaze breaker, will not remove much cylinder wall material. However, the finish it applies may only be adequate for certain types of rings. A rigid hone will remove more cylinder wall material during use, but the honing stones can provide a more precise finish and crosshatch.

A rigid hone is always used when a cylinder is bored. If during the inspection process it was determined that honing was all that was required, a minimum amount of cylinder wall material should be removed to ensure that the correct piston-to-wall clearance is maintained. In this instance, a flexible hone is usually the best choice, although a rigid hone can be used if used carefully.

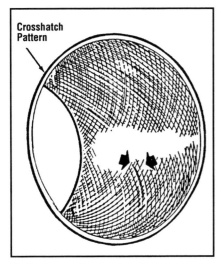

Cylinder wall crosshatch.

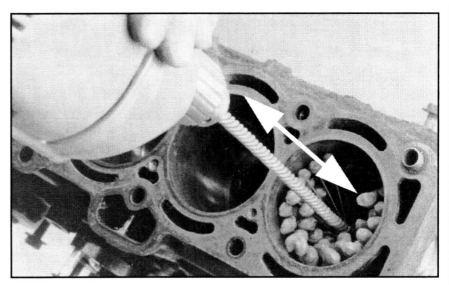

Honing a cylinder with a flexible hone.

Using a torque plate can reduce bore distortion when honing cylinders. The torque plate simulates the loading that occurs when the cylinder head is installed on the block. *(Courtesy: Perfect Circle/Dana Corp.)*

When the machinist bores the block, he will leave about 0.0025 - 0.003-in. (0.064mm - 0.076mm) cylinder wall material for finish honing. He will then use a multistep honing procedure, beginning with roughing stones and progressing to finishing stones when he is within 0.0005-in. (0.0127mm) of final bore size. In general, 220 grit stones are used for finishing when cast iron or chrome rings are used. 280 to 400 grit stones should be used for moly rings. To be sure of proper ring sealing, always consult with the machinist for the specific engine and/or piston ring manufacturers surface finish recommendations.

If you are using a rigid hone to apply a surface finish to a cylinder that did not require boring, attach the hone to a suitable drill, lubricate the honing stones with cutting oil and position the hone in the cylinder. Tension the stones, then turn on the drill motor and move the hone up and down in the bore to produce a crosshatch on the cylinder walls. Most ring manufacturers call for a crosshatch of 45 degrees to 60 degrees (included angle between the grooves) when measured at the center of the cylinder, or an angle of 25 degrees to 45 degrees when the grooves are measured from horizontal. Remove only as much cylinder wall material as necessary to leave a finish. Check the bore size frequently with a dial bore gauge.

Many of today's thin-wall blocks have reduced rigidity. The use of torque plates (also called honing plates or deck plates) is recommended on some engines to keep bore distortion to a minimum when honing. Torque plates are heavy iron plates that simulate cylinder heads when bolted to the block. The plate duplicates the bore distortion that occurs when the head bolts are tightened. Consult with a machinist to see whether torque plates are required.

The flexible hone is used in the same manner as the rigid hone, but of course without any stones to adjust. The crosshatch left by a flexible hone will usually not be as defined as that created by a rigid hone.

ENGINE ASSEMBLY

Preparation

When all machine work has been completed, the block and all components should be thoroughly cleaned and the piston and connecting rods assembled, if necessary. All component parts should be cleaned, laid out and prepared for assembly.

All threaded holes in the cylinder block should be cleaned and repaired, the block deburred, and thoroughly cleaned with soap and water.

Inspect all threaded holes, clean them with a thread chaser and re-

TRAINING FOR CERTIFICATION

pair as necessary. Clamping force measured when reading a torque wrench changes with friction. When the threads and the underside of the hex head on the bolt are not clean, friction will increase the torque reading.

Damaged threads can be repaired by drilling the hole oversize and cutting new threads with a special tap. A prewound insert is then installed with a special tool.

Major damage to threaded holes is repaired by drilling out the stripped threads, tapping the hole and installing an insert. Some prewound threaded inserts require special size drills and taps as well as a special installation tool, all of which is usually included in a kit. There are also solid type threaded inserts which use standard size drills and taps and are installed using a screwdriver that fits into slots on the insert.

Inspect the block for sharp edges and casting burrs and flash, paying special attention to the cam gallery, oil drain passages, main bearing bulkheads and lifter valley. Sharp edges are stress risers that can give cracks a starting point. If not removed, casting burrs and flash could break off later and cause engine damage. Large casting burrs and flash can be carefully chipped off with a hammer and chisel. The rest, along with the sharp edges, can be ground off with a high speed grinder and carbide cutter.

WARNING: Be sure to wear eye protection when deburring and cleaning the block.

Thoroughly wash the block with soap and hot water. Half of all ring failures that occur during break-in can be directly attributed to improperly cleaned cylinder walls. Following the cylinder honing or deglazing process, the cylinder walls must be cleaned with hot, soapy water and a stiff nonmetallic bristle brush. Gasoline, kerosene or other petroleum solvents cannot wash away abrasive residues. Rather, they carry the abrasives further into the pores of the metal where they can escape later.

All the grit, dirt and machining chips accumulated during the machining process must be removed from the block. A selection of stiff bristle brushes of various diameters should be used to dislodge dirt, grit and chips from all the nooks and crannies in the block, as well as the oil passages and lifter bores. Wash the cylinder bores last, after the rest of the block has been cleaned, to prevent scratching the cylinder wall finish.

Note: *Throughout the block washing process, be careful not to let any machined surfaces rust. The cylinder bores especially, will quickly rust after they have been scrubbed clean if a rust inhibitor such as WD-40® is not quickly applied.*

Once the block has been completely cleaned and rinsed, and all machined surfaces have been protected with a rust inhibitor, blow dry the block with compressed air, paying particular attention to the oil galleries. Do not use shop towels for drying. Lint from the towels can accumulate and cause problems. When the block is clean and dry to your satisfaction, and all machined surfaces have been protected, cover the block with a plastic trash bag if it is not to be assembled immediately.

The crankshaft should be washed with clean solvent. Run a stiff bristle brush through all the oil passages to make sure they are clean. After the crankshaft has been cleaned, blow it dry with compressed air, paying special attention to the oil passages.

Wash the pistons and connecting rods with clean solvent, dry them and assemble as required. Be sure to properly position the piston and rod prior to assembly, so that they will be facing the proper direction when installed in the engine.

On pistons with floating wrist pins, lubricate the wrist pin, connecting rod small end and piston bore, and assemble using new pin retainer clips. Some retainer clips will have a rounded edge and a sharp edge. These clips should always be installed so that the rounded edge is against the pin, leaving the sharp edge to bite into the piston aluminum. The clips should also be positioned with their gaps facing down, so that the full area of the clip is against the groove during combustion.

Pistons with pressed pins are assembled in either of two ways, by pressing the pin into the connecting rod or by heating the small end of the rod enough that

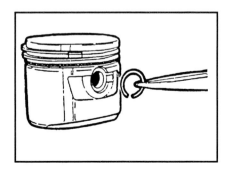

Always position wrist pin clips with the gap facing down.

Pressing the piston pin into the connecting rod small end bore.

Installing camshaft bearings in a cylinder block.

the pin can be installed using hand pressure.

Camshaft Bearings

Camshaft bearings are installed using tools that either drive or draw the bearings into place. Care must be taken to properly align the oil holes in the cam bearings with those in the block so oil can be pumped to the cam journals. If a hole is partially or totally blocked, it will result in cam and/or bearing failure. If the engine has bearings for balance shaft(s) or an auxiliary shaft, install them at this time.

After the camshaft (and auxiliary shaft) bearings are in place, install the rear camshaft (and auxiliary shaft) plug(s). Apply the recommended sealer or anaerobic adhesive to the plug to prevent oil leaks, then stake it in place to aid retention. If an oil gallery plug comes out, all oil pressure to the engine will be lost. This is usually an internal leak (within the timing cover).

Oil gallery plugs on the outside of the engine are most often threaded. Install these next, along with core plugs and wear sleeves. Coat core plugs with sealer or anaerobic adhesive and carefully tap them into place. Stake the edges of the plugs to assure that they cannot come out. Put sealer or teflon tape on threaded gallery plugs to prevent leaks and then install them.

CAUTION: If teflon tape is used, be absolutely certain that a piece of tape does not extend into the oil gallery. A loose piece could come off and block the flow of oil. Teflon also acts as a thread lubricant. When tightening a tapered pipe plug, do not overtighten it.

Crankshaft

Make sure the bearing saddles in the block are perfectly clean and dry because even the smallest amount of dirt will decrease bearing clearance. Dirt can also result in deformation in the bearing shell, enough to ruin a bearing.

Install the upper main bearing inserts (the ones with the oil holes in them) in the engine block. Next, install the lower inserts in the main caps. In some bearing sets, lower main bearings will not have lubrication holes in them. In other sets, both the upper and lower bearing halves have holes. Check to see that the oil holes in the upper bearing shells line up with those in the block. Bearing locating lugs must also be correctly positioned in their slots.

When the engine has a split rear main oil seal, install the seal halves in the block and rear main cap. Line up the surfaces where the seal halves meet so that they are above and below the parting half of the main cap. This will ensure that oil cannot leak through.

When the engine uses a one-piece rear main oil seal, it can be installed after the crankshaft and rear main bearing cap have been installed. Some engines have a rear seal housing that is bolted to the back of the block. When the seal fits into a recess in the rear bearing cap, it is often possible to install a full round seal around the crankshaft sealing flange before installing the rear main cap. The seal is held in place while installing the rear main cap. If the cap is already in place, a special installation tool may be used for seal installation.

Main bearing oil clearance can be checked using a dial bore gauge or with Plastigage™. If a dial bore gauge is to be used, the

TRAINING FOR CERTIFICATION

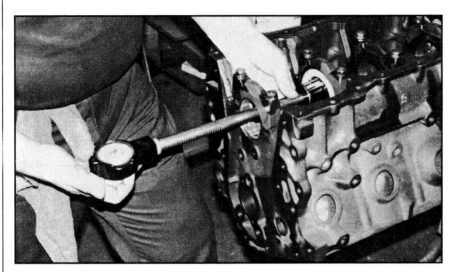
Measuring crankshaft main bearing oil clearance with a dial bore gauge.

main bearing caps should be installed and the bolts torqued to specification. Measure the crankshaft bearing journal diameter with a micrometer and use the measurement to set the dial bore gauge. Measure the inside diameter of the bearings using the bore gauge; the difference of this measurement and the main journal diameter measurement will be the main bearing oil clearance.

If bearing clearance is to be checked with Plastigage™, carefully lower the crankshaft onto the upper main bearings. Lay a piece of Plastigage™ lengthwise on the crank journal. Install and torque the main cap to specifications. Be careful not to turn the crankshaft while the Plastigage™ is in position. Remove the main cap and compare the thickness of the crushed piece of Plastigage™ to the gauge on the package to determine the amount of bearing clearance.

Refer to the manufacturer's recommendations for bearing clearance. A ballpark figure that applies to most engines is 0.001-in. (0.025mm) of clearance for each inch of journal diameter.

After checking bearing clearances, remove the crankshaft. Plastigage™ is easily removed with engine oil because it is soluble in oil. Lubricate the main bearings and crankshaft main journals with assembly lubricant.

Reinstall the crankshaft and main bearing caps. Lightly tap each main bearing cap with a hammer to seat it in its recess in the block. Then torque the main cap bolts to specifications. When the engine has upper and lower thrust halves, leave the thrust main until the last. Pry the crankshaft forward and backward to align the thrust halves, before tightening the main cap.

An anaerobic sealer or RTV silicone is sometimes installed on rear main caps to prevent oil leaks. Be careful not to get any under the parting surface of the main cap, as this will increase bearing clearance. Anti-seize compound is used with main cap bolts that are threaded into aluminum blocks.

Check crankshaft end-play by inserting a feeler gauge between the rear main bearing flange and crank, or by placing a dial indicator against the end or nose of the crank. End-play can then be measured by pushing the crank fore-and-aft as far as it will go. Refer to the manufacturer's recommended specifications.

Pistons And Connecting Rods

Piston Ring Installation

Place a new piston ring into the cylinder in which it will be installed and measure the ring end gap clearance with a feeler gauge. If the bore is worn, the ring should be positioned in the area of ring travel that is least worn. Square the ring in the cylinder using an inverted piston. The ring end gap should be within the range specified by the manufacturer. As a rule, the minimum end gap will be 0.003-in. - 0.004-in. (0.076mm - 0.102mm) per inch of cylinder diameter. Some manufacturers specify even less. An end gap that is too large is not as critical as one that is too small. A typical passenger car can have up to 0.030-in. (0.762mm) more end gap than the minimum specification without suffering from excessive blow-by. However, if an end gap is too small, as the ring heats up during engine operation, the ring will expand and the ends of the ring can butt together. As the ring continues to expand, the ring face will be forced against the cylinder wall causing rapid wear. If

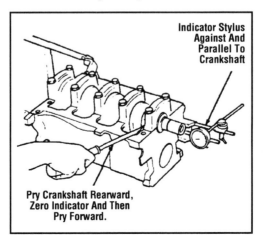
Measuring crankshaft end-play with a dial indicator. *(Courtesy: Ford Motor Co.)*

Measuring piston ring end gap.

the gap is too small, a rotary ring file is used to carefully trim the ends of a ring to bring it within specification. Use a fine whetstone to debur the edges of the ring after filing.

After the ring end gaps are set, install the rings on the pistons. Locate the side of the ring that is marked 'up'. This mark is usually a dot, however, refer to the information on the ring packaging for specific instructions. Use a ring expander to install the rings on the piston. Do not over-expand the rings or spiral them on by hand. Doing so can distort the rings, leaving a permanent twist that prevents the rings from seating properly.

Install the oil ring first. The middle compression ring is next, followed by the top compression ring. The position of the rings on the pistons is staggered so the end gaps are at least 60 degrees and preferably 180 degrees apart. This prevents combustion gases from having a direct route to blow by the rings. Follow the recommendations of the manufacturer.

Piston And Connecting Rod Installation

Make sure the connecting rod bearing saddles are clean and dry. Install the bearing inserts, making sure the locating lugs are correctly positioned in their slots. Some connecting rods have oil squirt holes built into the rod. Make sure the hole in the bearing lines up with the hole in the rod.

Lightly lubricate the piston, piston rings and cylinder bore. Many pistons are slightly offset or have valve recesses, depressions or raised domes which must face one way. A small notch or mark usually indicates which side of the piston faces the front of the engine. The front is usually the end that the harmonic balancer mounts on, whether the engine is mounted in the usual north/south configuration or transversely (as in many front-wheel drive cars).

Before installing the pistons and rods, install rod bolt protectors or pieces of fuel hose over the ends of the rod bolts to protect the crank journals. Position the crankshaft so the rod journal is beneath the cylinder in the bottom dead center position. A ring compressor is used to compress the rings into their grooves while the piston is slid into the cylinder. A soft hammer or wooden handle can be used to gently push the piston into place.

Lay a piece of Plastigage™ lengthwise on the crank journal. Install and torque the rod bearing cap to specifications. Be careful not to turn the crankshaft while the Plastigage™ is in position. Remove the rod cap and compare the thickness of the crushed piece of Plastigage™ to the gauge on the package to determine the amount of bearing clearance.

Remove the Plastigage™ from the crankshaft, then push the piston and rod assembly back up into the bore just enough to lubricate the upper rod bearing and crankshaft with assembly lube. Pull the rod back onto the crankshaft. Lubricate the lower

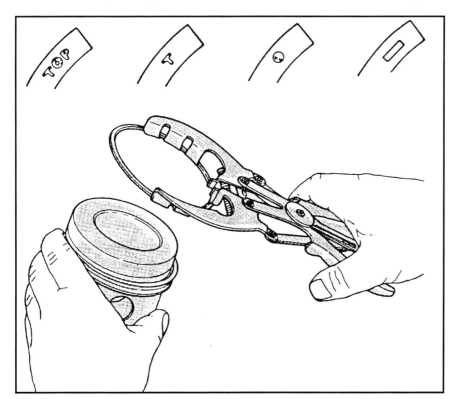

Installing piston rings with a ring expander. Piston rings are usually marked so the correct side can be installed facing up.

TRAINING FOR CERTIFICATION

Using a ring compressor to install a piston into the cylinder block.

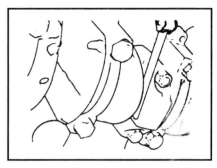

Check connecting rod side clearance with a feeler gauge.

rod bearing and crankshaft with assembly lube, install the rod bearing cap and torque the nuts to specifications.

Check connecting rod side clearances with a feeler gauge. There is seldom a problem with this measurement when the original crankshaft and connecting rods are used. However, if the crank has been welded or replaced, or any rods have been replaced, clearance can be below minimum specification. Clearance that is too tight can result in the rod bearings riding on the journal radius and failing. If clearance is not adequate, the rod can be machined to provide the necessary clearance.

Camshaft And Timing Components

Before sliding the camshaft and any auxiliary shafts into the block, coat the lobes with suitable assembly lubricant. Be very careful while sliding the cam into the block as the cam lobes can nick or damage the soft cam bearings. Install long bolts into the camshaft sprocket mounting holes or temporarily install the sprocket to provide leverage during installation.

Do not force the cam too far in, as it may push the rear cam plug out. The cam lobes can also be damaged in a careless installation. Rotate the cam to assure that it turns freely. Binding indicates improper clearances, misalignment in the cam bores or a bent camshaft.

Note: Sometimes the weight of the block hanging from a universal engine stand is enough to distort it. If the camshaft cannot be turned easily or will not go into its bores, remove the block from the stand and support it on a work-

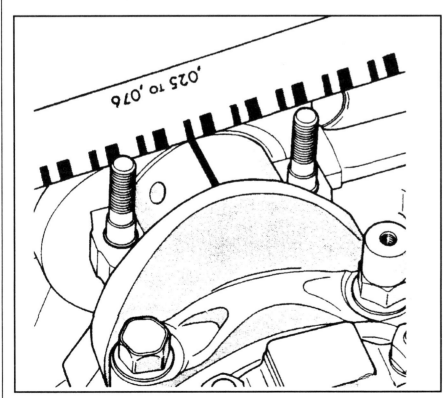

Measuring connecting rod bearing clearance with Plastigage™.
(Courtesy: DaimlerChrysler Corp.)

bench. This is often enough to allow the camshaft to turn easily.

Some engines have a thrust plate and retainer on the front of the camshaft. Use a feeler gauge or micrometer to check the clearance between the camshaft boss (on the back of the gear) and thrust plate. The spacer will be about 0.002-in. (0.051mm) thicker than the boss on the gear. This provides the necessary end-play. Change the spacer as needed to achieve the required clearance.

After installing the camshaft on an engine with a thrust spacer, end-play can be checked with a dial indicator against the nose of the cam. Push the cam as far in as it will go and set the gauge to zero. Then pull it as far out as it will go and note the reading on the gauge.

Timing gears come as a matched set and must likewise be replaced as a set. When installing a new timing belt, replacement of its sprockets usually are not necessary unless the sprockets show signs of unusual wear or are damaged.

Correctly position the timing marks on the camshaft and crankshaft gears or sprockets. Before installing a timing chain,

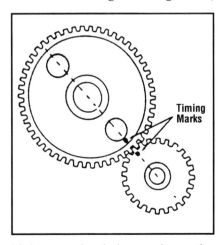

Make sure the timing marks on the cam and crank gears or sprockets (and any auxiliary or balance shaft sprockets) are correctly aligned.

lubricate it with clean engine oil. Torque the bolt(s) holding the cam gear in place to manufacturer's specifications.

When installing a timing belt, rotate the crankshaft until the number one piston is at TDC. Verify the positions of the camshaft and crankshaft timing marks. Install the new belt and adjust the tension.

Many engines with timing belts have spring loaded tensioners that provide an automatic tension adjustment. A lock bolt maintains the tension once the belt is in place. Timing belts stretch very little. Once tension is set, it will be maintained for the life of the belt. A new belt might not be exactly centered on the sprockets, however. After initially tensioning the belt, rotate the crankshaft to align the belt on the sprockets. Loosen the lock bolt and retighten it to provide accurate tension.

Tensioners that require manual adjustment must be set according to the manufacturer's instructions. Too much tension is bad for the belt and/or cam bearings. Too little tension may allow timing variations or belt slippage. Slippage will take the teeth off of the belt.

On an OHC engine, if the cylinder head or block deck has been milled, cam timing will be affected. Removal of 0.020-in. (0.508mm) from the head surface will retard cam timing by about one degree. This can be compensated for by installing a shim available from gasket manufacturers.

Proper timing of the balance shaft(s) in engines so equipped is critical. The offset weights on the balance shaft(s) are positioned to cancel out engine vibrations. The shafts are usually driven by a crankshaft mounted gear and rotate opposite to the direction of crankshaft rotation. If a shaft is incorrectly timed, it will amplify, rather than diminish engine vibrations.

Gasket Installation

When installing valve covers, timing covers, oil pans and other covers, never reuse old gaskets, like molded silicone rubber gaskets for instance. These are often found on late model passenger car engines because they last longer than cork rubber gaskets, are more flexible, conform better and seal tighter. Oil makes silicone swell, which keeps the gasket tight in use. But because it swells, molded silicone gaskets will usually not fit back in place.

Synthetic rubber gaskets must be installed against a perfectly clean and dry surface. Do not use gasket sealer on any synthetic rubber gaskets. It acts like a lubricant, causing the gasket to slip and leak.

Compression is what prevents a gasket from leaking. A cork/rubber gasket is compressed to about 50% of its original height. With silicone rubber, 30% compression is desirable. Less compression results in a poor seal, while more compression risks splitting the gasket. Unlike cork/rubber, which has many little air pockets in it, silicone is solid and does not compress. It deforms to either side to seal the joint. For that reason, both surfaces and the gasket must be clean and dry. Otherwise the gasket can slip.

Small metal grommets are often used in the gasket bolt holes to prevent the gasket from being over-tightened. The thickness of the grommet is such that the ideal amount of clamping force is achieved when the flange bottoms against the grommet.

Cork/rubber gaskets can be substituted for molded silicone rubber gaskets on some engines. But with a die cast cover, the flange design may make it impossible to use anything but an identical replacement.

Here are some gasket installation tips to keep in mind:

TRAINING FOR CERTIFICATION

- Lip-type oil seals must be lubricated to protect them against damage during initial start-up
- Do not over tighten the cover bolts. This can crush or damage the gasket and/or cover flange
- If using RTV silicone instead of a cut or molded gasket to seal a cover or oil pan flange, make sure the flange and engine surfaces are clean, dry, and oil free. RTV silicone sealer can only be used on covers with flat flanges, not those with raised lips or beads. Apply a 1/8-in. (3.175mm) to 3/16-in. (4.762mm) wide bead of sealer along the length of the cover flange or block surface, going to the inside or all the way around bolt holes. Be careful not to smear the bead of sealer as the cover is being installed.

Harmonic Balancer

One often overlooked item is the harmonic balancer or vibration damper on the front of the crankshaft. A defective, loose or incorrect vibration damper can result in a broken crankshaft. The rubber ring upon which the damper weight is mounted sometimes deteriorates. If the weight has shifted position, false ignition timing readings will be given.

Some harmonic balancers are a press fit on the crankshaft; however, they should never be hammered on. If the balancer is hammered on, not only will the balancer be damaged, but the crankshaft thrust bearing may be damaged as well. Installation tools are available that thread into the end of the crankshaft. The balancer is then started onto the crankshaft; a large bearing and nut are threaded onto the tool and then used to press the balancer into position.

FLYWHEEL CHECKS

On externally balanced engines, the flywheel must be correctly indexed when it is mounted on the crankshaft to maintain engine balance (many flywheels have bolt holes that are unevenly spaced so they can only go on one way). Make certain that index marks are properly aligned when installing the flywheel. Torque flywheel bolts to specifications in a crisscross pattern in several steps.

Next, check runout on the face of the flywheel with a dial indicator. If runout exceeds specifications, the flywheel may be warped or the crankshaft flange may be bent. Dirt or a burr between the flywheel and crankshaft flange can also cause this problem.

If the engine is for a vehicle with a manual transmission, check the condition of the pilot bushing or bearing in the end of the crankshaft. Install a new one if necessary.

When there is a separate transmission bell housing, its alignment can be checked with a dial indicator. The customer complaint for transmission to engine misalignment would be clutch chattering or hard shifting. First, check to make sure the circular transmission opening in the bell housing is centered with respect to the crankshaft. Then, check to see that the machined face of the bell housing is square (parallel) to the back of the engine block.

To check for concentric runout in the circular transmission opening, position a dial indicator on the back of the flywheel so its tip rides against the inside edge of the opening. Rotate the crankshaft and note the amount of runout. If it exceeds specifications, usually about 0.008-in. (0.203mm) maximum, the housing can be re-centered on the block using offset dowel rods.

To check the bell housing-to-block alignment, mount the dial indicator on the flywheel so its tip rides on the outside machined face of the opening in the bell housing. Again rotate the crankshaft and note the amount of runout. If the reading varies by more than the manufacturer's specifications, shims can be installed between the bell housing and block at the required bolt positions to restore proper alignment.

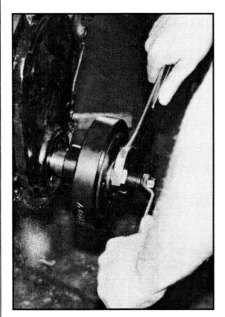

Always use a proper installation tool to install a harmonic balancer.

NOTES

NOTES

NOTES

TRAINING FOR CERTIFICATION

LUBRICATION AND COOLING SYSTEMS DIAGNOSIS AND REPAIR

LUBRICATION SYSTEM

Oil Pressure Testing

Make sure the engine oil is at the proper level. Remove the oil pressure sending unit and install an accurate oil pressure gauge. Start the engine and observe the oil pressure reading. Refer to the manufacturer's specifications for the temperature and engine speed at which oil pressure should be checked.

The following are possible causes of lower than specified oil pressure:
- Incorrect oil viscosity or diluted oil
- Clogged oil filter
- Stuck oil pressure relief valve
- Worn oil pump
- Loose or damaged oil pump pickup
- Clogged pickup screen
- Excessive bearing clearance
- Cracked, porous or clogged oil gallery
- Missing oil gallery plug.

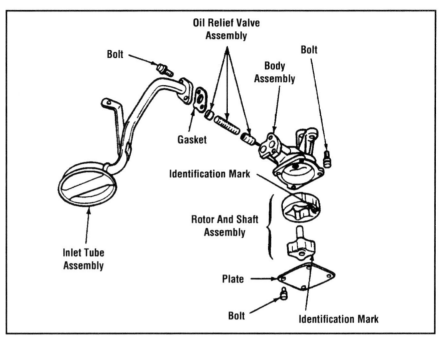

Oil pump exploded view. *(Courtesy: Ford Motor Co.)*

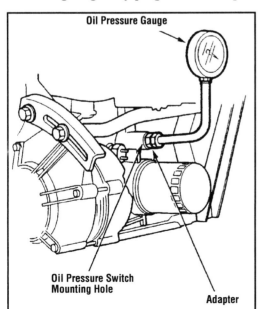

Checking oil pressure with an oil pressure gauge. *(Courtesy: Honda Motor Co.)*

Oil Pump

Oil pumps that are enclosed within the oil pan are usually replaced during a rebuild. Where the oil pump housing is part of another engine component like the timing cover, the oil pump is rebuilt.

Remove the oil pump cover and seal, if equipped. Remove the gears or rotors from the pump body. Mark the gears prior to removal so that they can be reinstalled with the same gears meshing. Remove the pressure relief valve pin and the piston and spring. Remove and discard the oil pump pickup.

Clean all parts in solvent. Check the inside of the pump housing, the pump cover and the gears or rotors for nicks, burrs, scoring, damage or excessive wear. Replace parts or the entire pump, as necessary.

Check the pump cover surface with a straightedge and feeler gauge. Measure the rotor or gear thickness and diameter with a micrometer. Place the rotors or

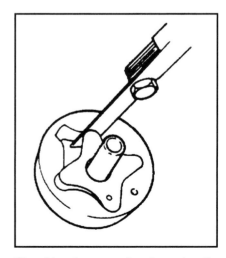

Checking inner and outer rotor tip clearance.
(Courtesy: Ford Motor Co.)

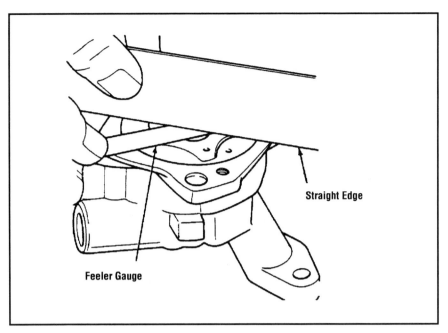

Checking rotor end-play. *(Courtesy: Ford Motor Co.)*

gears in the pump housing and measure the clearance between the two with a feeler gauge. Place a straightedge across the pump housing and measure the gear or rotor end-play with a feeler gauge. Compare your measurements with the manufacturer's specifications.

Measure the relief valve spring's free length or tension and compare to specification. Check the relief valve piston for scoring and for free movement in the bore.

Lubricate the pump housing and gears or rotors with clean engine oil. Install the rotors or gears in the pump housing. Be sure to install gears with the matchmarks aligned. Install the pump cover with a new seal, if equipped, and torque the bolts to specification. Install the pressure relief valve assembly and retain it with the pin. Install a new pickup and screen, making sure it is at the proper angle to fit closely to the bottom of the oil pan.

Prior to installation, prime the pump by pouring clean engine oil into the pickup and operating the drive gear or rotor assembly to draw oil through the pump.

Oil Cooler

Some vehicles are equipped with an engine oil cooler. There are oil-to-air coolers and oil coolers that are incorporated into the vehicle's cooling system. Inspect the cooler, lines, hoses and connections of air-to-oil coolers for leaks. Make sure the cooling fins are not bent or damaged.

If oil is found in the cooling system or coolant in the oil, on vehicles with liquid cooled oil coolers, the oil cooler assembly should be inspected for leakage.

COOLING SYSTEM

Proper engine operation depends on the cooling system maintaining engine operating temperature within a specified range. If the engine runs colder than normal, a richer air/fuel mixture may result, causing poor fuel mileage and excessive emissions. If the engine runs hotter than normal, it may cause detonation or the engine may overheat, resulting in possible engine damage.

Cooling System Inspection

Begin cooling system inspection by checking the coolant level, then check coolant concentration using an antifreeze tester. The protection level should be at least −20°F (−30°C). Visually check the engine and cooling system components for leaks and damaged components. Replace any hoses that show signs of leakage or are swelling, chafing, hardened or have soft spots. Always use new clamps when replacing a hose.

Check the radiator cap rating to make sure it is the right one for the vehicle. Check the cap's relief valve spring action, and inspect the seal for brittleness. Check the filler neck on the radiator or surge tank mating surface.

Check the water pump drive belt for wear, glazing and belt tension. A slipping belt will not turn the pump impeller at the proper speed to circulate coolant. If the engine is equipped with a mechanical fan, a slipping belt will cause the fan to turn too slowly, not draw enough air through the radiator, and possibly cause the engine to overheat. Replace or adjust the belt as necessary.

Inspect the fan for missing, cracked or bent blades. If equipped with a fan clutch, check the back of the clutch for an oily film, which would indicate that fluid is leaking and replacement is necessary. Turn the fan and clutch assembly by hand; there should be some viscous drag, but it should turn smoothly during a full rotation. Replace the fan clutch if it does not turn smoothly or if it does not turn at all. It should also be replaced if there is no viscous drag when hot or cold.

If the fan is electric, make sure it runs when the engine warms up and also when the A/C is switched on. Make sure the fan shroud is in place and not broken.

Start the engine and listen for unusual noises. A hammering

TRAINING FOR CERTIFICATION

sound may indicate a restriction in the water jacket or air in the system. Squealing noises indicate a bad belt or water pump bearing damage. Gurgling from the radiator may point to air in the system.

Cooling System Pressure Testing

Use a hand-held pump with a pressure gauge that is designed for cooling system testing. While the engine is cold, remove the pressure cap from the radiator or surge tank. Make sure the system is filled to capacity, then attach the tester. Pump it up to the rated system pressure and watch the gauge needle; it should not drop rapidly. If pressure drops, check for leaks at the radiator and heater hoses, water pump, radiator, intake manifold, sensor fittings, water control valves and heater core. Repair leaks as required and retest.

If you can't spot the leak, it may be internal such as a head gasket, cracked cylinder head or cracked block. Inspect the engine oil for signs of coolant; if it is thick and milky, that's a dead giveaway. Start the engine and watch the tester gauge. If the pressure immediately increases, there could be a head gasket leak, but not into the crankcase. The coolant may be going out the tailpipe, however, which would be indicated by white smoke from the exhaust pipe and a somewhat sweet antifreeze odor in the exhaust. Remember that catalytic converters can mask small coolant leak symptoms, because the converter super-heats the coolant into a fine vapor that is not noticeable.

Use the cap adapter to check the pressure cap. Pump it up to the cap's rating. It should hold for about 10 seconds and then decrease gradually. If it drops too quickly, or if it does not drop at all when the pressure test exceeds the cap's rating by 1-3 psi, replace the cap.

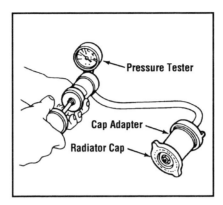

When pressure testing the radiator cap, the proper cap adapter must be used. The cap should be capable of holding the pressure recommended for the vehicle. Pressure test the cooling system at approximately the release pressure of the radiator cap.

Drive Belts, Pulleys And Tensioners

Check the accessory drive belt(s) for evidence of cracking, fraying, glazing or other damage and replace as necessary.

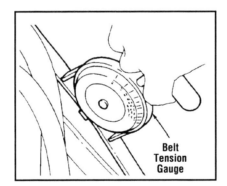

Checking belt tension with a belt tension gauge.
(Courtesy: Ford Motor Co.)

If the belt is adjustable, belt tension can be checked using the deflection method or by using a belt tension gauge. Locate a point midway between the longest accessible belt span. If using the deflection method, push on the belt with your finger using moderate pressure and measure the belt deflection. If you are using a belt

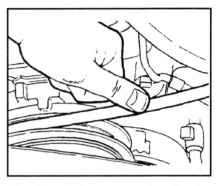

Checking belt tension using the deflection method.

tension gauge, position the gauge and measure the amount of force necessary to deflect the belt. Compare your reading with specifications.

Belt tension should also be checked on vehicles with automatic belt tensioners to make sure the tensioner is functioning properly. Some automatic tensioners are equipped with belt length indicator and minimum and maximum acceptable marks, the theory being that if the correct length belt is installed on the engine and the mark is within range, belt tension is correct.

To adjust V-belt tension, loosen the adjuster pulley, or accessory pivot and adjuster bolts. Then use a suitable prybar to move the pulley or accessory until the belt tension is correct. Tighten all fasteners and recheck belt tension.

If belt replacement is necessary, loosen the adjuster pulley or accessory pivot and adjuster bolts, moving the pulley or accessory to eliminate belt tension, and remove the belt. It may be necessary to remove other accessory drive belts to gain access to a particular belt.

Before removing a serpentine V-ribbed belt, make sure there is a belt routing diagram handy or draw one prior to belt removal to prevent installation problems. Use a socket or wrench to tilt the automatic tensioner away from the belt, and then remove the belt

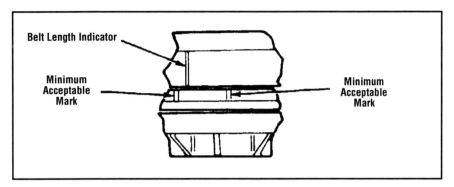

Belt length indicator on an automatic belt tensioner.
(Courtesy: Ford Motor Co.)

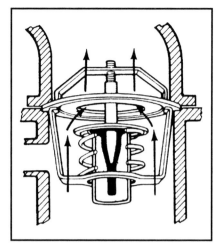

The thermostat is designed to open when the engine coolant reaches a specified temperature.

from the pulleys.

After the belt is removed, spin the pulleys to determine if they wobble or if an accessory has any noticeable bearing wear. Inspect the pulleys for chips, nicks, cracks, tool marks, bent sidewalls, severe corrosion or other damage. Check for hard objects such as small stones or sand that may have become imbedded in the bottom of the pulley grooves.

When replacing the belt, inspect all pulleys for improper alignment. Aligned pulleys reduce both pulley and belt wear, and vibration of engine components. If the belt pulleys are severely misaligned, look for improper positioning of an accessory or its corresponding pulley, improper fit of the pulley or shaft, or incorrect components installed.

Install a new belt, making sure it is correctly positioned in its pulley grooves and properly routed. Adjust the belt tension, as required.

Thermostat

The thermostat's function is to allow the engine to come to operating temperature quickly and then maintain a minimum operating temperature. If the thermostat is good, the upper radiator hose should be hot to the touch after the engine has been idling and warm. If the hose is not hot, the thermostat is most likely stuck open, especially if there has also been a complaint of poor heater performance. If the thermostat was stuck closed, the engine would quickly overheat.

To test the thermostat operation, remove the radiator cap while the engine is cold. Put a thermometer in the radiator fill neck and start the engine. Keep an eye on the coolant in the radiator and occasionally feel the upper radiator hose. When the hose gets warm, the coolant should be moving in the radiator. Check the thermometer. If it doesn't go above 150°F (66°C), the thermostat is either stuck open or missing.

The thermostat can be checked more precisely by removing it from the vehicle and submerging it a pan of water with a thermometer. Heat the water and observe the temperature when the thermostat opens. The opening temperature should match the thermostat's temperature rating. Remove the thermostat from the water. The valve should close slowly when exposed to cooler ambient temperature.

To replace the thermostat, drain the cooling system and disconnect the hose from the thermostat housing. Remove the thermostat housing and remove the thermostat. Clean all gasket material from the sealing surfaces and check the surfaces for nicks or burrs.

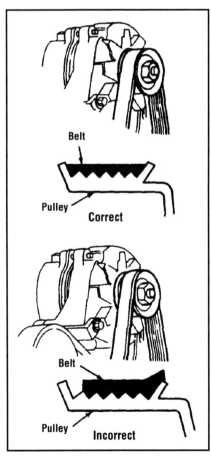

Make sure V-ribbed belts are properly seated in the pulley grooves. One revolution of the engine with the belt incorrectly seated can damage the belt.
(Courtesy: Ford Motor Co.)

TRAINING FOR CERTIFICATION

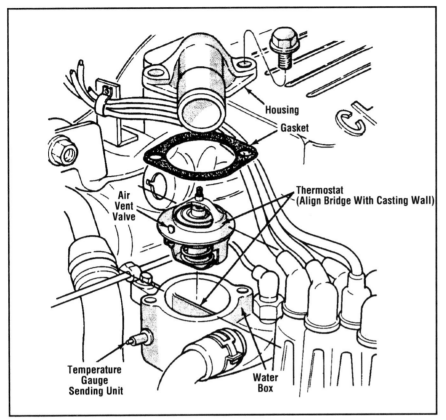

Typical thermostat installation. *(Courtesy: DaimlerChrysler Corp.)*

tom of the pump. Check the water pump bearings by grasping the fan or pulley and attempt to move the impeller shaft back-and-forth. If there is any movement, the water pump bearings are defective. Replace the pump if it leaks or the bearings are defective.

To replace a water pump, drain the cooling system and disconnect the hoses from the pump. Remove the water pump drive belt and pulley. Remove any brackets or other components necessary for water pump removal. Remove the water pump mounting bolts and remove the water pump. Clean all gasket material from the sealing surfaces and check the surfaces for nicks or burrs.

Install the water pump using new gaskets and torque the mounting bolts to specification. Install any brackets or other components that were removed. Install the water pump pulley and the drive belt. Properly tension the belt, as required. Connect the coolant hoses to the water pump. Refill the cooling system, start the engine and check for leaks.

Properly seat the thermostat in its flange on the engine. Make sure the heat sensing portion of the thermostat is installed so as to expose it to the hot coolant side. Using a new gasket, install the thermostat housing and torque the bolts to specification. Refill the cooling system, start the engine and check for leaks and proper operation.

Water Pump

The water pump is mounted on the engine and driven by a belt. The pump employs an impeller fan designed to pump coolant throughout the engine via specially placed water jackets. While some water pumps are mounted directly behind the radiator fan and its pulley, others are mounted independently on the front of the engine.

Check for a coolant leak at the water pump drain hole at the bot-

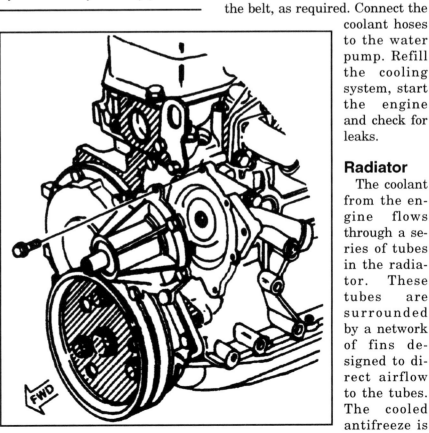

Typical water pump installation. *(Courtesy: GM Corp.)*

Radiator

The coolant from the engine flows through a series of tubes in the radiator. These tubes are surrounded by a network of fins designed to direct airflow to the tubes. The cooled antifreeze is then circu-

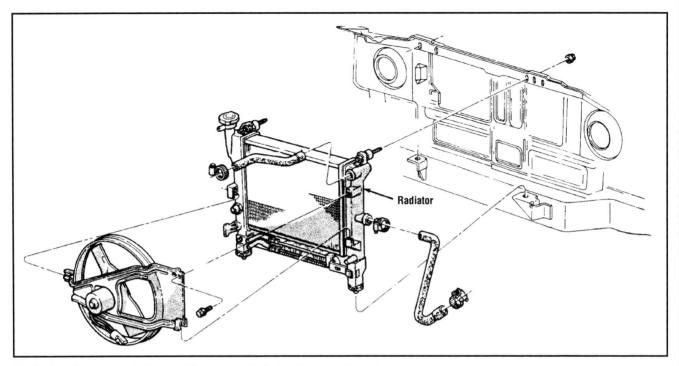

Typical radiator installation. *(Courtesy: DaimlerChrysler Corp.)*

lated back through the engine in order to maintain proper operating temperature.

Clean the radiator fins of debris, bugs or leaves that may have been drawn in while driving. Make sure all fins are intact, and not bent so as to misdirect airflow. Distorted fins can be straightened using a suitable tool. However, be careful when straightening because the fins are very delicate.

Inspect the radiator for damage and any signs of leakage from the core tubes, radiator tanks and hose collars. If visual inspection and/or pressure testing indicates that radiator replacement is required, drain the cooling system and disconnect the hoses and transmission cooler lines, if equipped. Separate the radiator from the fan shroud and electric cooling fan, if equipped, and remove the radiator mounting fasteners. Remove the radiator from the vehicle.

Transfer fittings and/or temperature sending units to the replacement radiator, as required. Position the radiator in the vehicle and tighten the mounting fasteners to specification. Install the shroud and electric cooling fan, if equipped. Connect the radiator hoses and transmission cooler lines, if equipped. Refill the cooling system, start the engine and check for leaks.

Flushing, Filling And Bleeding The Cooling System

Engine overheating can be caused by a clogged cooling system. If you suspect that the system is clogged, flush the system using one of the commercially available flushing kits. Flush the system according to the directions supplied with the kit.

After flushing, or whenever the cooling system has been drained for service, the system should be refilled with the vehicle manufacturers specified coolant mixture. Ethylene glycol is the most widely used substance mixed with water to form engine coolant, although propylene glycol and some other substances are becoming popular. All used coolants, however, pick up heavy metals such as lead from the solder used in the assembly of heat exchangers, and must either be disposed of properly or recycled.

A mix of 50 percent water/50 percent antifreeze is usually the most effective mixture, but some vehicle manufacturers may specify otherwise. This mixture lowers the freezing point while raising the boiling point, affording protection against boiling over. Also, a high quality antifreeze will prevent corrosion, rust formation and foaming. Drain and replace the coolant at the vehicle manufacturer's recommended interval, as the protective additives, such as the rust inhibitors, diminish in use.

Some vehicles now come factory equipped with long life coolant that can last up to five years. Long life coolant is usually identifiable by being a different color than the familiar yellow/green of conventional antifreeze. It is usually orange or red in color, but not

TRAINING FOR CERTIFICATION

all coolants that are different in color than yellow/green are necessarily long life. Always check to be sure. Also, although it is not necessarily harmful to the engine or other cooling system components to do so, conventional coolant and long life coolant should not be mixed. This can cut back on the life expectancy of the long life coolant, and also slightly diminish some of its protective properties.

When filling the cooling system, be aware that some vehicles have bleeder valves to release air trapped in the system and require that a specific bleeding procedure be followed. If there are no bleeders, to prevent air from becoming trapped in the system, you can remove a hose from the highest point (usually at the heater core) and fill the system until coolant begins to come out at this point. Start the engine, and as soon as the level in the radiator drops, top it off and install the cap. Fill the reservoir to the indicated level.

Most all cooling systems in use today are equipped with a coolant reservoir. The radiator cap in the system functions as a two-way check valve: It has a limit that, once exceeded, allows the coolant to escape from the radiator and into the reservoir. This happens normally as the coolant heats up and expands. As the system cools, it creates a partial vacuum and sucks coolant from the reservoir or surge tank back into the radiator. It is a closed system where, ideally, no coolant escapes and no air gets into the cooling system. Air in the system contributes to corrosion.

When adding coolant to this type of system, add it to the reservoir, noting that the tank is usually marked with the proper coolant level. The overflow tank serves as a receptacle for coolant forced out of the radiator overflow pipe and provides for its return to the system. As the engine cools, the balancing of pressures causes the coolant to siphon back into the radiator.

Cooling Fans

Mechanical Cooling Fan

Most rear-wheel drive cars and trucks have belt-driven, mechanical fans equipped with a fan clutch. The fan clutch is designed to slip when cold and rotate the fan at certain maximum speeds when hot. Fan clutches improve gas mileage and reduce noise levels.

Inspect the fan clutch as described under Cooling System Inspection in this study guide. To test the fan clutch, attach a thermometer or electronic temperature probe to the radiator near the inlet and connect a timing light to the engine. Start the engine and strobe the fan to 'freeze' the blades; note the engine speed. When the engine warms up (check the thermometer), the fan speed should increase and the blades will appear to be moving in the strobe light. As the temperature drops, the fan should slow down. A quick check of fan clutches is to shut down a hot engine, then observe how long it takes for the fan to stop spinning. A properly operating clutch should stop the fan from spinning within two seconds.

Replace the fan clutch if it fails inspection or testing.

Electric Cooling Fan

Many vehicles, especially those with transverse engines, use electric cooling fans. Besides not needing a belt to drive them, electric fans conserve energy since they run only when needed.

When the engine slightly exceeds proper operating temperature, the electric fan should come on. It may cycle on and off as the coolant warms and cools. On most vehicles, the fan also should run whenever the A/C is switched on. (Some vehicles have two fans with one dedicated to the A/C system. That one may not run for engine cooling alone.)

If the fan doesn't run, check for

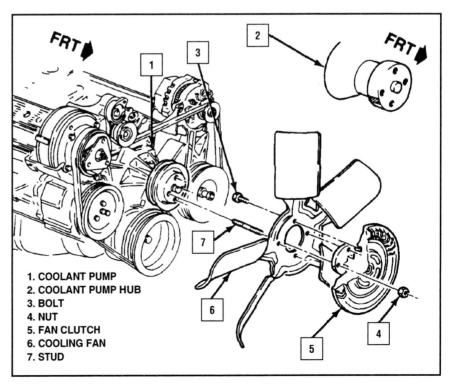

1. COOLANT PUMP
2. COOLANT PUMP HUB
3. BOLT
4. NUT
5. FAN CLUTCH
6. COOLING FAN
7. STUD

Typical mechanical cooling fan and fan clutch. *(Courtesy: GM Corp.)*

power at its connector using a test light. If there is power, the fan motor is faulty. If there is no power, the problem may be a blown fuse, a bad relay, the fan's temperature switch, the engine coolant temperature sensor, the computer controls or wiring. In most cases, the fan motor can be replaced by removing the entire assembly, and then removing the fan from its motor. Use caution not to damage the radiator fins or core tubes when removing the fan.

NOTES

NOTES

TRAINING FOR CERTIFICATION

FUEL, ELECTRICAL, IGNITION AND EXHAUST SYSTEMS INSPECTION AND SERVICE

FUEL AND AIR INDUCTION SYSTEM

Inspect fuel system components for leaks and damage. Replace carburetor, fuel pump or throttle body gaskets, as necessary. Check fuel lines, hoses and connections for cracks, leaks, torn O-rings or other damage. Inspect fuel system wiring for chafing and damaged connectors.

Clean the intake manifold gasket mating surfaces. Inspect the manifold for cracks or other damage. Check the surface that is mounted to the cylinder head for warpage using a straightedge and feeler gauge and compare to manufacturer's specifications.

Install the intake manifold using a new gasket(s). Apply sealer as recommended by the vehicle and gasket manufacturers. Tighten the intake manifold bolts to the specified torque in the proper sequence.

Remove the air filter from its housing and gently tap it on a hard surface to dislodge dirt. Hold the filter up to a light source to check the filter. If the filter is clean, light will pass through all areas. Dark areas on a filter element will not allow light to pass through; the filter is dirty and must be replaced.

When replacing the filter element, wipe any dirt from the housing and make sure the filter seals fit properly. Check and replace the crankcase breather filter. Check the air intake ducting for cracks and poor joint sealing that could allow air into the engine other than through the filter. This is especially important on vehicles with remote mounted MAF (Mass Airflow) or VAF (Vane Airflow) sensors. These sensors measure the amount of air entering the engine and the PCM (Powertrain Control Module) uses this information in calculating the proper air/fuel mixture. Air entering the throttle body other than through the MAF or VAF sensor will not be accounted for in these calculations.

POSITIVE CRANKCASE VENTILATION

The PCV (Positive Crankcase Ventilation) system vents crankcase gases into the engine air intake where they are burned with the air/fuel mixture. The PCV system keeps pollutants from being released into the atmosphere, and also helps to keep the engine oil clean by ridding the crankcase of moisture and corrosive fumes.

Remove the PCV valve from its mounting and shake it. Listen or feel for the rattle of the valve plunger within the valve body. If it rattles, the valve is not stuck open or closed. If it doesn't rattle, the valve must be cleaned or replaced.

Reinstall the PCV valve and connect a tachometer to the engine. With the engine at normal operating temperature and idling, pinch off the vacuum hose between the PCV valve and intake manifold. The engine speed should drop 50-60 rpm if the PCV system is operating properly. If not, look for a blockage in the system, clean or repair as necessary and retest.

You should always feel a strong suction at the PCV inlet when the engine is idling. If you don't feel that suction, look for a collapsed PCV hose, a clogged PCV port in the intake manifold or a PCV valve that's stuck closed. Should the PCV valve stick open or its hose crack open, the engine will idle rough and manifold vacuum will be low.

Check for vacuum through the whole system by removing the fresh air supply hose from the air cleaner and placing a stiff piece of paper over the hose end. Wait one minute; if vacuum holds the paper in place, the system is OK.

Remember that an inoperative PCV system can allow engine blow-by to pressurize the crankcase and either create oil leaks or make existing leaks worse. Check the general condition of the PCV

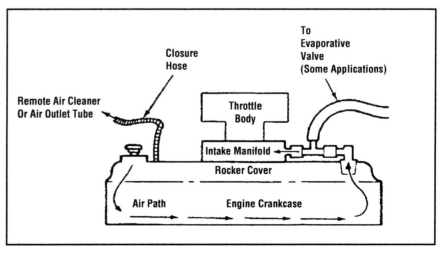

PCV system functional diagram. *(Courtesy: Ford Motor Co.)*

components, making certain that the valve, the breather cap, filter, tubes, orifices and hoses are serviceable.

TURBOCHARGERS AND SUPERCHARGERS

Turbochargers and Superchargers are used to increase engine power by forcing more air into the cylinders.

Turbochargers

A turbocharger is an air pump divided into two sections, the turbine and the compressor. The turbine is attached to the exhaust manifold where a turbine wheel inside the turbine housing is driven by the exhaust gas pressure and heat energy. The turbine wheel is connected by a shaft to the compressor wheel inside the compressor housing. The spinning of the turbine wheel causes the compressor wheel to spin, drawing in air to the compressor housing where it is compressed and pumped through ducts into the intake manifold. As the speed of the turbine increases, so does the pressure output, or boost, of the compressor.

Boost pressure must be limited to prevent engine damage. Boost is controlled by a wastegate or by shutting off the fuel supply to the engine.

A wastegate is a valve activated either by a diaphragm or a boost control solenoid. Wastegates are either integral to the turbine housing or are remotely mounted in the exhaust system. If controlled by a diaphragm, when a preset boost limit is reached, the diaphragm moves a rod that opens the wastegate. If controlled by a boost control solenoid, which is operated by the PCM, the wastegate opens and closes in response to sensor inputs to the PCM. When the wastegate is opened, excess exhaust pressure is released from the turbine hous-

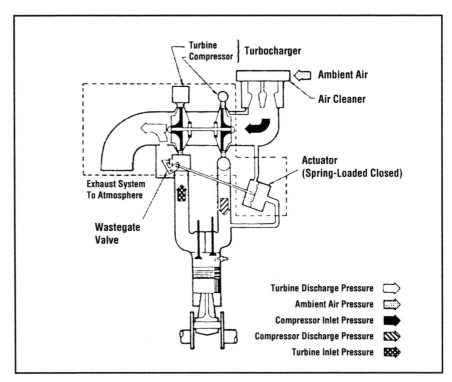

Turbocharger operation. *(Courtesy: GM Corp.)*

ing, directed to the exhaust system and expelled into the atmosphere.

The fuel supply to the engine can be shut off by the PCM in response to inputs regarding intake manifold pressure or engine speed. The MAP (Manifold Absolute Pressure) sensor sends a signal to the PCM when a specified intake manifold pressure is reached. The PCM then cuts the fuel supply to the engine, causing boost and engine speed to decrease. When intake manifold pressure falls below the limit, fuel delivery resumes. When boost is controlled in response to engine speed, the PCM will cut the fuel supply when inputs are received that a specific engine speed has been reached. Fuel delivery resumes when engine speed drops below the limit.

Turbocharger related engine performance problems are caused by too little boost pressure or by overboost. These problems can usually be traced to a malfunction in the boost control system. However, other components should also be inspected before any are condemned. If there is a lack of power, check for a dirty air cleaner, loose or restricted intake ducting or restricted exhaust system. Listen for unusual noises coming from the turbocharger that could be an indication that the rotating assembly is binding or dragging.

A wastegate actuator that is stuck can be the cause of too little boost and low power if it is stuck open, or overboost if it is stuck closed. Check for free movement of the actuator by hand if possible, and check for obstructions that could prevent free movement or closure. Wastegate operation can be checked using air pressure and a pressure gauge. Consult the appropriate service manual for testing procedures and pressure specifications.

WARNING: Turbochargers operate at extremely high tem-

TRAINING FOR CERTIFICATION

peratures. Do not touch the turbocharger while the engine is operating. Allow the turbocharger to cool sufficiently after the engine has been turned off before performing testing or servicing procedures.

Most turbocharger failures are caused by lubrication problems such as oil lag, restriction or lack of oil flow and foreign material in the oil. The exhaust flow past the turbine wheel creates extremely high temperatures, which creates a harsh operating environment for the turbocharger shaft bearings. Some manufacturers connect coolant lines to the turbocharger to cool the shaft bearings, but others rely on engine oil to lubricate and to cool. With the latter design, it is a good idea to let the engine idle for about a minute before shutting it off, particularly if the vehicle has been run hard, to let oil cool the turbocharger. If the engine is shut off immediately, the oil may burn, causing hard carbon particles to form. This in turn, will destroy the bearings.

The oil and filter should be changed at regular intervals and the air filter should be inspected regularly. Inspect the routing and integrity of the oil supply and oil drain lines and check for oil leaks. When a turbocharger is replaced, it should be preoiled prior to installation and the engine should not be revved before proper oil pressure has been established.

Superchargers

A supercharger is similar to a turbocharger in that it acts as an air pump. However, a supercharger is driven by a belt from the engine crankshaft, instead of by the exhaust gases. The amount of boost pressure generated is determined by the size of the pulleys on the crankshaft and supercharger, and by engine speed. The size of the pulleys can be varied in order to get the desired boost. It may be necessary to drive the supercharger at a speed faster or slower than crankshaft speed. If the supercharger turns the same speed as the engine, it is driven 1:1. If it's geared to turn faster for more or quicker boost, it is overdriven. If it's set up to run slower than the engine, it is under driven. Otherwise, boost increases as engine speed increases.

To prevent supercharger cavitation, reduced performance and increased temperatures, a bypass valve is installed at the supercharger outlet. This bypass valve allows a controlled amount of air flow from the supercharger outlet back into the supercharger.

Proper supercharger performance depends on there being no vacuum leaks, which could cause a lean operating condition. Vacuum leaks can be detected using a propane cylinder.

EXHAUST SYSTEM

Inspect the exhaust system for leaks which might cause excessive noise, or more importantly, a safety hazard by allowing noxious exhaust fumes inside the vehicle. Check the pipes, muffler(s) and catalytic converter(s) for corrosion and damage. Check the hangers for wear, cracks and hardening. Check the heat shields for corrosion and damage. All connections throughout the system should be secure.

A restricted exhaust system will cause a lack of power and poor fuel economy. To check for a restricted exhaust system, connect a vacuum gauge to the intake manifold. At idle, there should be about 17-21 in. Hg vacuum. Accelerate the engine gradually to 2000 rpm. The vacuum should momentarily drop to zero and then return to normal without delay; if the exhaust is restricted, as the engine rpm is increased the vacuum will slowly drop to zero and slowly rise to normal. When closing the throttle, the vacuum should momentarily increase and then resume the normal reading; if the exhaust is restricted the vacuum will not increase when the throttle is closed. Accelerate the engine to 2500 rpm and hold. If the vacuum reading drops 3 in. Hg below the original reading

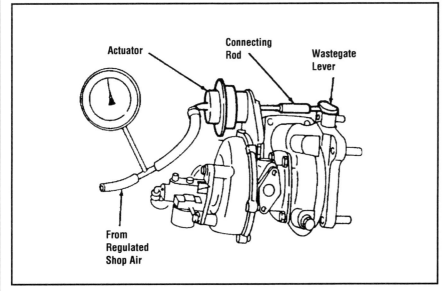

Testing wastegate operation using air pressure.
(Courtesy: Ford Motor Co.)

after a few minutes, the exhaust is restricted.

When inspecting the exhaust system for restrictions, check for obvious causes like dents or kinks in the system. Some vehicles use double wall tubing for exhaust pipes. The inside tube can collapse or rust inside the outer tube and cause a restriction, even though the outer tube looks OK. Tap on the exhaust pipes with a mallet and listen for rattling or rust breaking loose which would indicate a problem inside the pipe. A clogged catalytic converter can also cause a restriction.

BATTERY

Preliminary Inspection

WARNING: The sulfuric acid in battery electrolyte can cause serious injury if it contacts the eyes or skin. To prevent injury, always wear skin and eye protection when servicing the battery. Batteries give off hydrogen gas, which is highly explosive. Never smoke or allow flames near a battery.

Visually inspect the battery, looking for damage to the battery case and damage or corrosion on the battery terminals and cables. If the battery case is damaged and there is any evidence of leakage, the battery must be replaced. Check the battery's date of manufacture. Just because the battery is near the end of its service life does not mean that it will necessarily test bad. However, the age of the battery must be considered when deciding whether replacement is necessary.

Corrosion on the battery case, and battery tray and hold-down, can be cleaned with a solution of baking soda and water. Make sure the battery tray is in good condition and the battery is mounted securely without overtightening the battery hold-down.

If the battery terminals and cables are corroded, remove the cables, negative cable first, and clean the terminals and cables with a battery brush. Before disconnecting the battery cables, keep in mind that computers, programmable radios, and other solid-state memory units may have their memories erased by disconnecting the battery. In addition, the engine and transmission on some vehicles may perform erratically when first started and must undergo a relearning process once the battery is reconnected. To prevent this, a 12-volt power supply from a dry cell battery can be connected to the cigarette lighter or power point connector to maintain voltage in the system while the battery is disconnected.

Inspect the entire length of the battery cables for heavy corrosion, frayed wires and damaged insulation, and replace as necessary. Secure the cables to the battery terminals after cleaning, and apply a coating of petroleum jelly to the terminals to minimize further corrosion.

If the battery has removable vent caps, check the electrolyte condition and level in each battery cell. Look for cloudy or muddy discoloration of the fluid. Discolored fluid is a sign of recent deep cycle discharge action. Add distilled water to the proper level, if necessary. In general, the electrolyte should be 1/4 - 1/2-in. (6.35-12.70mm) above the plates.

Battery State-Of-Charge

A hydrometer is used to check the specific gravity of the battery's electrolyte. A hydrometer is a transparent glass tube with a rubber pickup hose on one end and a rubber bulb on the other end. Inside the tube is a weighted float which is scale-calibrated to read specific gravity in a range of about 1.100 to 1.300.

The floats in the hydrometer are calibrated with 80°F (27°C) being the exact reference point. For each 10°F variation above or below the 80°F (27°C) mark, 0.004 specific gravity points are added (for temperatures above 80°F) or subtracted (for temperatures below 80°F).

A battery with a specific gravity reading of 1.260 (usually stated as twelve-sixty) is generally regarded as fully charged, while a battery with a reading of 1.070 (ten-seventy) is generally regarded as fully discharged. It follows that 1.120 is one-quarter charged; 1.170 is half-charged; and 1.215 is three-quarters charged. A maximum of 0.050 (50 points) difference between cells is all that is allowed. Any difference greater than this calls for further testing of the battery, and the battery may have to be removed from service.

When checking specific gravity, hold the hydrometer in a vertical position and insert the draw tube into the battery. Draw just enough electrolyte into the tube to permit the float to move freely

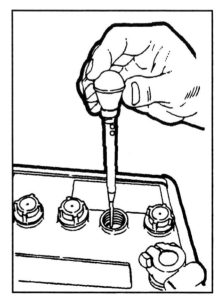

Checking electrolyte specific gravity with a hydrometer.

TRAINING FOR CERTIFICATION

without touching the top, bottom or sides. Note that there are times when electrolyte readings are inaccurate. One is just after adding water to the battery. After adding water, it is recommended to wait at least one day, or until the vehicle has been operated for a while to check specific gravity. Another time is during or just after charging. After charging, it is recommended to wait at least 15 minutes prior to checking specific gravity. Still another time is just after the battery has been subjected to a high rate of discharge, such as after prolonged cranking.

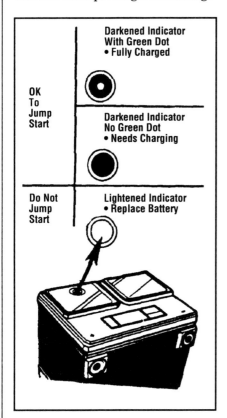

Charge indicator in a sealed maintenance-free battery.

The electrolyte cannot be checked on sealed maintenance-free batteries. Instead, these batteries have a built-in charge indicator. Usually a good battery is indicated by a green or light-colored dot in the center. If the indicator is dark, the battery may be jumped or recharged. If the indicator is clear or light yellow, the fluid level is below the level of the hydrometer; the battery should not be charged and should be replaced.

Check the open circuit voltage of the battery to determine if it must be charged before further testing. Remove the surface charge from the battery by turning on the high beams for 10 seconds, then wait a few seconds before checking the battery voltage. Make sure all accessories are off. Then measure the voltage at the battery terminals with a voltmeter. If the voltage is less than 12.4 volts, the battery must be charged before proceeding with further testing.

Battery Capacity Testing

Connect the battery load tester to the battery terminals. Turn the load control knob to draw current equal to three times the ampere-hour (amp/hr) rating or one-half the CCA (Cold Cranking Amps) rating. The battery rating is usually given on the battery case, case top or label, and some labels will indicate the load that should be placed on the battery. Maintain the load for 15 seconds, then check the voltage reading. On a good battery, the voltage should be 9.6 volts or higher, however, the voltage may be slightly lower if the ambient temperature is less than 70°F (21°C).

If a battery fails a load test, but was deemed OK for testing when the state-of-charge was checked, connect a voltmeter and battery charger to the battery and charge the battery for three minutes with the charger set at 40 amps. If after three minutes the voltage reading is greater than 15.5 volts, replace the battery.

STARTER

System Diagnosis

The most common starter complaints are that the starter does not crank or cranks too slowly to start the engine, the starter operates but does not turn over the engine, and finally, that the starter stays engaged once the engine has started.

Engine Does Not Crank

If nothing happens when the key is turned to the Start position, turn the headlights on and turn the key to Start again. If the headlights do not dim, check for an open circuit in the starting system. If the lights go dim, test the battery as described in the previous section of this study guide. If the battery tests OK, check the condition of the cables and wiring to the starter and solenoid, making sure all connections are clean and tight.

If the wiring connections are OK, next check the starter solenoid function. Connect a jumper wire between the battery and solenoid 'S' terminals. If the solenoid clicks, but the engine still doesn't crank, perform the starter current draw test. Make sure that the starter is mounted securely and that the starter drive and flywheel/flexplate ring gear are not binding. Perform the voltage drop test. If current draw is excessive and voltage drop is within specifications, the starter is defective or there is a mechanical problem that is preventing the engine from turning over. If the engine can be rotated by hand through several revolutions, replace the starter.

Engine Cranks Slowly

If the engine cranks slowly, test the battery as described in the previous section of this study guide. If the battery tests OK, check the condition of the cables and wiring to the starter and solenoid, making sure all connections are clean and tight. If the wiring connections are OK, perform the starter current draw test.

Make sure that the starter is mounted securely and that the starter drive and flywheel/flexplate ring gear are not binding. Perform the voltage drop test. If current draw is excessive and voltage drop is within specification, the starter is defective or there is a mechanical problem that is causing the engine to bind. If the engine can be rotated by hand through several revolutions, replace the starter.

Starter Operates But Does Not Crank Engine

If the starter turns but the engine does not, the starter drive is not engaging the flywheel/flexplate ring gear properly. You may hear a high-pitched grinding sound when this happens. If possible, have an assistant operate the ignition key while you check starter drive operation. Look for a bent or warped flywheel/flexplate, missing ring gear teeth or a damaged or failed starter drive.

Make sure the starter is mounted securely. If engagement is poor and the flywheel/flexplate and starter drive are OK, the position of the starter motor may allow adjustment if equipped with shims, otherwise the starter should be replaced.

Starter Does Not Disengage

If the engine starts and the starter does not disengage, check for a damaged starter drive. Check the starter drive/ring gear engagement. If engagement is poor, the position of the starter motor may allow adjustment if equipped with shims, otherwise the starter should be replaced.

If the starter drive is working and engaging properly, but still does not disengage when the engine starts, check for a defective solenoid or a short in the control circuit.

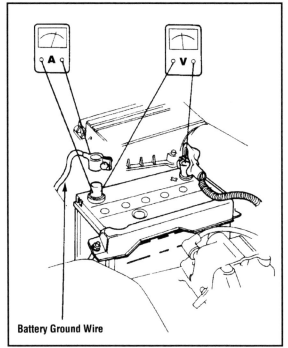

Performing a starter current draw test. *(Courtesy: Honda Motor Co.)*

Starter Current Draw Test

Perform a starter current-draw test with a high-amperage-rated ammeter designed for this purpose. Allow the engine to warm up to normal operating temperature, then disable the ignition system on gasoline engines or fuel system on diesel engines. Connect the ammeter's inductive pickup to the negative battery cable or connect the ammeter in series with the negative battery cable. Also, connect a voltmeter across the battery terminals.

With the test connections made and all electrical loads off, crank the engine over with the starter. Starter current draw when cranking the engine over should be around 150 amps for a 4-cylinder engine, 200 amps for a 6-cylinder and 250 amps for a V8. However, some permanent magnet type and gear reduction starters can draw more. Always refer to the specifications in the vehicle service manual. Battery voltage should remain above 9.6V, and the engine should spin rapidly.

Higher than normal current-draw readings may be caused by a short circuit in the starter motor or mechanical problems causing binding in the engine. Amperage that is too low may be caused by starter circuit resistance.

Voltage Drop Test

Connect the voltmeter black lead to the battery cable connection at the starter. Make sure the test point is positioned after the starter relay. Now, disable the engine ignition and crank the starter motor. As you crank the engine, touch the red lead of the voltmeter to the positive post of the battery. Quickly switch the voltage scale knob of the voltmeter to progressively lower scales until a reading can be seen. Be sure to remove the red voltmeter lead from the battery post before you release the starter engagement or you could damage the meter.

If the voltage drop exceeds 0.5V for the entire starter circuit, repeat the test, moving the black lead toward the battery one connection at a time. When you get past the bad one, the voltage drop reading will sharply decline. Once the high resistance connection is identified, clean the connection or replace the part causing the problem. Refer to the proper service manual for the manufacturer's maximum acceptable voltage drop specifications.

Starter Ground And Connection

To check the starter ground circuit, first place the black lead of the voltmeter on the negative battery post. Now, using the same

TRAINING FOR CERTIFICATION

technique you used on the supply voltage circuit, crank the starter and place the red lead on the starter housing. Read the voltage drop. If you find a high resistance at the starter mounting, the presence of engine-to-starter shims or an engine-to-battery ground cable that produces more than the voltage drop recommended by the manufacturer, check carefully for a poor connection or even corrosion.

Locate the problem by moving the red lead closer to the battery post in steps. Repeat the voltage test each time. Don't forget to remove the red lead before you stop cranking the engine!

Now check the control circuit. Locate the starter relay or solenoid and place the black voltmeter lead on the relay control voltage terminal. Crank the engine over and place the red lead to the positive battery terminal to check the voltage drop. This checks the circuit through the ignition switch, neutral safety switch and wiring harness.

Remember to check the relay ground. Quite often, relays are mounted on non-metallic inner fender panels and require a good ground to the metal vehicle body.

IGNITION SYSTEM

Inspect the distributor cap for cracks, wear or damage. Damaged or worn out distributor caps and rotors can cause a variety of problems, such as hard starting, hesitation, lack of power, rough idle, high emissions and poor fuel economy. A good visual inspection is usually all that is required for distributor caps. Always look for the obvious: corrosion build-up in the high tension wire terminals, cracks and worn tips under the cap and rotor contacts.

Remove the cap. Some or all of the plug wires might have to be unclipped or detached to remove the cap. If so, number the wires and the cap before removal.

Clean the cap inside and out with a clean rag. Check for carbon paths. A carbon path shows up as a dark powdery trail, usually from one of the cap sockets or inside terminals to a ground. Check for any fine cracks in the cap or rotor. A crack may allow moisture to enter and could cause voltage to arc around the inside of the cap. Check for eroded terminals in the cap or rotor. Check the carbon button in the top center of the distributor cap for damage or wear.

If any of the above conditions exist, replace the distributor cap and rotor.

Visually inspect spark plug wires for brittleness, cracking, burn marks and corrosion, especially at the distributor cap and spark plug terminals. Problems with spark plug wires and spark plugs are most easily seen on an engine analyzer or oscilloscope. Wires with excessively high resistance will be indicated by an abnormally high firing line on a scope pattern. The resistance of suspect wires can be checked using an ohmmeter.

Check spark plugs for appearance and spark plug gap. The spark plug electrodes should be squared off and the insulator should be a light tan. Check spark plug gap using a wire gauge and bend the side electrode

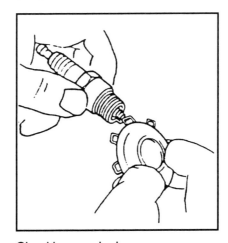

Checking spark plug gap.

to adjust to specification. A gap that is too tight will cause a lower firing voltage resulting in incomplete combustion. If the gap is too wide, the coil may not generate enough voltage to bridge the gap for the required time.

Ignition Timing

Ignition timing is a measurement in degrees of crankshaft rotation, of the point at which the spark plug fires in relation to when the piston reaches TDC on the compression stroke. Initial or basic timing is usually set for the spark plug to fire just before the piston reaches TDC, when the engine is at idle. However, as engine speed increases, the spark plugs must fire earlier to allow complete combustion, requiring a mechanism to advance spark timing.

On engines with distributorless ignition systems, basic ignition timing and timing advance are controlled by the PCM and are not adjustable. Modern distributor ignition systems allow adjustment of initial timing after disconnecting a timing control circuit to keep the PCM from advancing the timing during timing adjustment. Spark advance on these systems is controlled by the PCM.

On older distributor ignition systems, spark advance is controlled by a mechanical advance mechanism using centrifugal weights inside the distributor and by a vacuum advance unit. Initial timing can be adjusted on these systems after the vacuum hose is disconnected from the advance unit and plugged.

To adjust initial timing, connect a timing light and tachometer to the engine. Disconnect the vacuum hose or electrical connector, as required, to prevent spark advance. Run the engine at the specified rpm and aim the timing light at the timing marks on the crankshaft pulley or flywheel. If adjustment is required, loosen the distributor hold-down and turn the distributor housing until the timing marks align. Tighten the hold-down and recheck timing.

To check mechanical timing advance, accelerate the engine, check the advance using a timing light and compare to specifications. If spark advance is not within specification, inspect the mechanical advance mechanism for binding and clean and lubricate as necessary.

The vacuum advance unit can be checked using a timing light and a vacuum pump. Check the timing advance while applying vacuum to the advance unit with the pump. The advance unit should be rated at certain amounts of advance at specific vacuum increments. If the vacuum advance unit is not within specifications, it may require adjustment or replacement.

NOTES

NOTES

Prepare yourself for ASE testing with these questions on ENGINE REPAIR

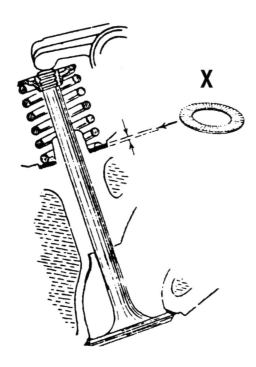

1. Technician A says that Part 'X' shown above is used to rotate the valve spring. Technician B says that Part 'X' is used to correct installed spring height. Who is correct?
    A. Technician A only
    B. Technician B only
    C. Both A and B
    D. Neither A or B

2. A compression test shows that one cylinder is too low. A leakage test on that cylinder shows that there is excessive leakage. During the test, air could be heard coming from the tail pipe. Which of the following could be the cause?
    A. broken piston rings
    B. bad head gasket
    C. bad exhaust gasket
    D. an exhaust valve not seating

3. Technician A says that main bearing oil clearance can be checked with Plastigage™. Technician B says that main bearing oil clearance can be checked with a dial bore gauge. Who is correct?
    A. Technician A only
    B. Technician B only
    C. Both A and B
    D. Neither A or B

4. Which of the following is NOT a method of correcting excessive valve stem to guide clearance?
    A. knurling the valve stem
    B. knurling the guide
    C. reaming for oversize
    D. guide replacement

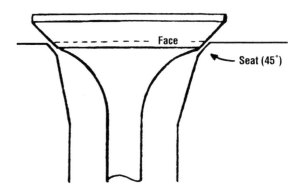

5. To raise the seat contact on the valve face shown above, which combination of stones should be used?
    A. 45 degree and 60 degree stones
    B. 30 degree and 45 degree stones
    C. 15 degree and 30 degree stones
    D. 30 degree and 60 degree stones

6. A vacuum gauge is connected to the intake manifold of an engine running at idle. The pointer on the gauge fluctuates rapidly but steadies as the engine speed is increased. The test results indicate:
    A. a leaking intake manifold gasket
    B. worn valve guides
    C. late ignition timing
    D. a weak valve spring

7. A technician takes a compression reading on a 4-cyl. engine and gets readings of 140, 135, 135 and 40 psi. Then he performs a wet compression test and the readings are almost the same as those in the first test. Technician A says that a burned valve could cause these readings. Technician B says that a broken piston ring could cause these readings. Who is correct?
    A. Technician A only
    B. Technician B only
    C. Both A and B
    D. Neither A or B

8. Technician A says that valve timing that is incorrect can cause high cylinder compression readings. Technician B says that late ignition timing can cause a high intake manifold vacuum reading. Who is correct?
    A. Technician A only
    B. Technician B only
    C. Both A and B
    D. Neither A or B

9. Technician A says that severe cylinder bore taper can be determined by measuring the cylinder with a gauge. Technician B says that severe cylinder bore taper can be found by performing a cylinder leakage test. Who is correct?
    A. Technician A only
    B. Technician B only
    C. Both A and B
    D. Neither A or B

10. Camshaft lift is defined as:
    A. the distance the valves are open
    B. the height of the lobes on the camshaft
    C. the time the valves stay closed
    D. the time the valves stay open

11. Technician A says that a burned valve could be the result of too little valve lash. Technician B says excessive valve lash could cause very fast camshaft wear. Who is correct?
    A. Technician A only
    B. Technician B only
    C. Both A and B
    D. Neither A or B

12. Technician A says that engine cylinders wear most at the top of the ring travel. Technician B says that the difference in cylinder bore measurements taken parallel and perpendicular to the crankshaft will indicate cylinder bore taper. Who is correct?
    A. Technician A only
    B. Technician B only
    C. Both A and B
    D. Neither A or B

13. A tapping noise can be heard coming from the upper part of an engine running at idle. Which of the following could be the cause?
    A. valve clearance out of adjustment
    B. worn rocker arm shaft
    C. loose exhaust manifold
    D. all of the above

14. An engine's upper and lower center main bearings show more wear than the ones toward either end of the crankshaft. Technician A says this wear pattern could be due to a bent crankshaft and replacing it with one that is straight will fix the problem. Technician B says that the cylinder block main bearing bores are misaligned and the problem can be fixed by align boring or honing the block. Who is correct?
    A. Technician A only
    B. Technician B only
    C. Both A and B
    D. Neither A or B

15. Blue-gray smoke comes from the exhaust of a vehicle during deceleration. Of the following, which cause is **LEAST** likely?
    A. worn valve guides
    B. broken valve seals
    C. worn piston rings
    D. clogged oil return passages

Engine Firing Order: 1-8-4-3-6-5-7-2

16. During a cylinder balance test on the engine shown above, cylinder No. 5 and cylinder No. 7 were both found to be weak. Technician A says a blown head gasket could be the cause. Technician B says an ignition problem could be the cause. Who is correct?
    A. Technician A only
    B. Technician B only
    C. Both A and B
    D. Neither A or B

Prepare yourself for ASE testing with these questions on ENGINE REPAIR

17. When a cylinder head is disassembled, the stem height on one valve is found to be too high. Which of the following can correct this?
    A. regrind the valve
    B. regrind the valve seat
    C. install a new valve seat
    D. install a shim

18. All of the following methods can be used to find cracks in aluminum heads **EXCEPT**:
    A. pressure test
    B. magnetic particle detection
    C. visual inspection
    D. dye penetrant

19. Technician A says that before installing an oil pump, it should be primed with clean engine oil. Technician B says that before installing an oil pump, the position of the pickup screen should be checked to make sure it fits closely to the bottom of the oil pan. Who is correct?
    A. Technician A only
    B. Technician B only
    C. Both A and B
    D. Neither A or B

20. Crankshaft journals should be measured for:
    A. diameter
    B. out-of-round
    C. taper
    D. all of the above

21. All of the following can cause low oil pressure **EXCEPT**:
    A. diluted engine oil
    B. worn pistons
    C. excessive bearing clearance
    D. clogged oil pump pickup

22. While conducting a cylinder leakage test, bubbles are seen in the coolant in the radiator. Technician A says a failed exhaust valve is the cause. Technician B says a failed intake valve is the cause. Who is correct?
    A. Technician A only
    B. Technician B only
    C. Both A and B
    D. Neither A or B

23. Technician A says that umbrella valve seals are installed in their operating position during cylinder head assembly. Technician B says that positive valve seals are positioned by the valve. Who is correct?
    A. Technician A only
    B. Technician B only
    C. Both A and B
    D. Neither A or B

24. Technician A says that excessive grinding of the valve face and seat allows the valve tips to move too far away from their spring seats. Technician B says grinding the stem tips helps to correct minor changes in valve spring installed height. Who is correct?
    A. Technician A only
    B. Technician B only
    C. Both A and B
    D. Neither A or B

25. Technician A says that excessive cylinder head or block deck surfacing on a pushrod engine can change the geometry between the valve stem tip and rocker arm. Technician B says that excessive grinding of the valve face and seat changes the geometry between the valve stem tip and rocker arm. Who is correct?
    A. Technician A only
    B. Technician B only
    C. Both A and B
    D. Neither A or B

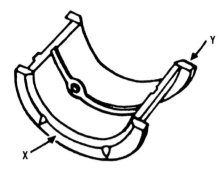

26. Technician A says that Plastigage™ is used to measure the clearance between the crankshaft and 'X' and 'Y' on the bearing shown above. Technician B says that a dial indicator is used to measure the clearance. Who is correct?
    A. Technician A only
    B. Technician B only
    C. Both A and B
    D. Neither A or B

27. Technician A says that piston rings installed upside down will cause increased oil consumption. Technician B says rings installed upside down will cause piston ring land damage. Who is correct?
    A. Technician A only
    B. Technician B only
    C. Both A and B
    D. Neither A or B

28. The diameter of the piston above is measured from point:
    A. 'A' to 'A'
    B. 'B' to 'B'
    C. 'C' to 'C'
    D. 'D' to 'D'

29. Of the following spring measurements, which would NOT require spring replacement if the measurement is not within specification?
    A. installed height
    B. pressure
    C. free height
    D. squareness

30. Technician A says that a hydraulic lifter rotates during operation. Technician B says the face of the lifter should be convex. Who is correct?
    A. Technician A only
    B. Technician B only
    C. Both A and B
    D. Neither A or B

31. Technician A says that valve lash should be adjusted with the piston in the cylinder being adjusted at TDC on the compression stroke. Technician B says that hydraulic lifters should be adjusted to zero lash. Who is correct?
    A. Technician A only
    B. Technician B only
    C. Both A and B
    D. Neither A or B

32. Technician A says that the thermostat controls maximum engine temperature. Technician B says if the engine continuously runs hot, install a cooler thermostat. Who is correct?
    A. Technician A only
    B. Technician B only
    C. Both A and B
    D. Neither A or B

33. Technician A says if the camshaft bores are more than 0.0015-in. (0.0381mm) out-of-round, they should be align bored or align honed. Technician B says that the block needs to be align bored or honed if bore alignment is not within 0.0015-in. (0.0381mm). Who is correct?
    A. Technician A only
    B. Technician B only
    C. Both A and B
    D. Neither A or B

Prepare yourself for ASE testing with these questions on ENGINE REPAIR

34. Technician A says that head bolts must be tightened in one step to achieve an accurate torque. Technician B says that torque-to-yield head bolts can be reused. Who is correct?
    A. Technician A only
    B. Technician B only
    C. Both A and B
    D. Neither A or B

35. Technician A is rebuilding an engine with a timing chain. Technician B is rebuilding an engine with a timing belt. Technician A says that the timing chain is stretched and must be replaced, however, the timing sprockets appear to be in good shape so they can be reused. Technician B says he is replacing the timing belt and that this requires him to also replace the timing sprockets. Who is correct?
    A. Technician A only
    B. Technician B only
    C. Both A and B
    D. Neither A or B

36. All of the following are ways to clean an aluminum engine **EXCEPT**:
    A. cold solvent tank
    B. carburetor cleaner
    C. hot caustic tank
    D. pyrolytic oven

37. All of the following are methods of repairing a cracked block **EXCEPT**:
    A. pinning
    B. magnafluxing
    C. gluing
    D. welding

38. In order to perform a proper compression test, all of the following conditions should be met **EXCEPT**:
    A. The throttle blades should be held wide open.
    B. The battery should be fully charged.
    C. The engine should be at operating temperature.
    D. Only the spark plug for the cylinder to be tested should be removed.

39. A single rod bearing shows excessive wear, but the rest show normal wear. Technician A says this could be due to an out-of-round connecting rod bore. Technician B says that this could be due to an obstruction in the crankshaft oil passage(s) that supplies the bearing. Who is correct?
    A. Technician A only
    B. Technician B only
    C. Both A and B
    D. Neither A or B

40. All of the following could cause an engine to ping or detonate **EXCEPT**:
    A. a faulty EGR valve
    B. retarded ignition timing
    C. excessive machining of the cylinder head
    D. using fuel with too low of an octane rating

41. Technician A says that a twisted connecting rod would cause a slanted wear pattern on the side of a piston. Technician B says that a bent connecting rod would cause the rod bearing inserts to have more wear on one side. Who is correct?
    A. Technician A only
    B. Technician B only
    C. Both A and B
    D. Neither A or B

42. An engine has a defective harmonic balancer. Technician A says this can cause inaccurate ignition timing readings. Technician B says this can cause a broken crankshaft. Who is correct?
    A. Technician A only
    B. Technician B only
    C. Both A and B
    D. Neither A or B

43. Technician A says the flywheel on an externally balanced engine can only be installed in one position on the crankshaft flange. Technician B says that flywheel runout should be checked with a straightedge and feeler gauge. Who is correct?
    A. Technician A only
    B. Technician B only
    C. Both A and B
    D. Neither A or B

44. All of the following could cause an engine to overheat **EXCEPT**:
    A. defective fan clutch
    B. clogged cooling system
    C. broken fan shroud
    D. stuck open thermostat

45. All of the following can cause an engine oil leak **EXCEPT**:
    A. cracked valve cover gasket
    B. stuck open PCV valve
    C. worn rear main seal
    D. worn piston rings

46. Technician A says that the engine should be accelerated when checking vacuum advance. Technician B says that the engine should be accelerated when checking centrifugal advance. Who is correct?
    A. Technician A only
    B. Technician B only
    C. Both A and B
    D. Neither A or B

47. Technician A says if a ring ridge is not removed, damage to the new piston and rings can result. Technician B says when the block will be rebored for oversize pistons, removing the ridge is not necessary. Who is correct?
    A. Technician A only
    B. Technician B only
    C. Both A and B
    D. Neither A nor B

48. All of the following could cause reduced performance from a turbocharged engine **EXCEPT**:
    A. dirty air filter
    B. restricted exhaust system
    C. worn turbocharger bearings
    D. stuck closed wastegate

49. All of the following statements about batteries are true **EXCEPT**:
    A. A fully charged battery should have a specific gravity of 1.260.
    B. If the open circuit voltage is less than 12.4 volts, charge the battery at a slow rate.
    C. A battery is load tested to one-third its CCA (Cold Cranking Amps) rating for 15 seconds.
    D. At the end of a load test, battery voltage must be 9.6 volts or more.

50. Technician A says that a deep gouge in a cylinder bore can be repaired by installing a sleeve. Technician B says a crack in a cylinder bore requires that the block be replaced. Who is correct?
    A. Technician A only
    B. Technician B only
    C. Both A and B
    D. Neither A or B

51. When a cylinder head with an overhead camshaft is discovered to be warped, which of the following is the **MOST** correct repair option?
    A. replace the head
    B. check for cracks, straighten the head, surface the head
    C. surface the head, then straighten it
    D. straighten the head, surface the head, check for cracks

52. Technician A says that crankshaft connecting rod journals typically experience the most wear at BDC (Bottom Dead Center). Technician B says that bearing journal diameter should be measured at the center and at both ends. Who is correct?
    A. Technician A only
    B. Technician B only
    C. Both A and B
    D. Neither A or B

Prepare yourself for ASE testing with these questions on ENGINE REPAIR

53. All of the following can result from resurfacing a cylinder block **EXCEPT**:
    A. change in the camshaft timing on an overhead camshaft engine
    B. change in the alignment between the cylinder heads and intake manifold on a V8 engine
    C. decreased compression
    D. change in the valvetrain geometry on a pushrod engine

54. Technician A says if only the upper front main bearing is worn, a belt adjusted too tight is indicated. Technician B says when an engine with a knock is disassembled, the connecting rod bearing(s) closest to the oil pump will show the most wear. Who is correct?
    A. Technician A only
    B. Technician B only
    C. Both A and B
    D. Neither A or B

55. A timing cover is being installed using a synthetic rubber gasket. Technician A says that a 1/8-in. (3.175mm) to 3/16-in. (4.762mm) wide bead of sealer should be applied along the length of the cover flange and block surface. Technician B says the gasket must be installed against a perfectly clean and dry surface. Who is correct?
    A. Technician A only
    B. Technician B only
    C. Both A and B
    D. Neither A or B

56. Technician A says that when replacing a thermostat, the heat sensing element should face the radiator. Technician B says that when checking a fan clutch, the fan should rotate easily. Who is correct?
    A. Technician A only
    B. Technician B only
    C. Both A and B
    D. Neither A or B

57. When cleaning a cylinder block for final assembly, which of the following should be used?
    A. carburetor cleaner
    B. kerosene or cutting lubricant
    C. hot, soapy water and a bristle brush
    D. petroleum solvent

58. Technician A says that a cylinder should be bored to within 0.001-in. (0.025mm) of final size. Technician B says that 220 grit stones should be used for finish honing when cast iron rings are used. Who is correct?
    A. Technician A only
    B. Technician B only
    C. Both A and B
    D. Neither A or B

59. Technician A says the retainer clips on pistons with floating pins should be reused if they are not damaged. Technician B says that after a full floating pin bushing is pressed into the rod, the piston and rod is then reassembled. Who is correct?
    A. Technician A only
    B. Technician B only
    C. Both A and B
    D. Neither A or B

60. All of the following statements about piston rings are true **EXCEPT**:
    A. Ring end gap must be measured prior to installation.
    B. Rings must be installed facing the correct vertical direction.
    C. Rings are installed on the piston using a ring compressor.
    D. Ring end gaps must be staggered around the piston.

61. A manifold vacuum test is performed on an engine in a shop that is located approximately 4000 ft. (1.2km) above sea level. With the engine at normal operating temperature, the gauge reads a steady 15 in. Hg at idle. Which of the following would be indicated by this vacuum reading?
    A. late ignition timing
    B. low compression
    C. vacuum leak
    D. normal operation

62. An engine has a light metallic knocking noise during light engine loads. However, when the cylinder with the noise is disabled during a cylinder balance test, the sound diminishes. Technician A says that the noise is caused by excessive connecting rod bearing clearance. Technician B says that the noise is caused by excessive piston-to-wall clearance. Who is correct?
    A. Technician A only
    B. Technician B only
    C. Both A and B
    D. Neither A or B

63. All of the following are statements describing normal mechanical fan clutch operation **EXCEPT**:
    A. A fan clutch has viscous drag regardless of temperature.
    B. A fan clutch varies fan speed according to engine temperature.
    C. A fan clutch stops the fan from spinning within two seconds after turning off a hot engine.
    D. A fan clutch varies fan speed according to engine speed.

64. Which of the following methods can be used to find a cylinder with an internal coolant leak?
    A. Draw vapors from the radiator into a vial containing a special chemical and look for a color change.
    B. Hold an exhaust analyzer probe over the radiator neck and check for the presence of exhaust gases.
    C. Perform a leakdown test and look for the presence of bubbles in the radiator coolant.
    D. all of the above

65. All of the following can cause an oil leak **EXCEPT**:
    A. worn piston rings
    B. worn valve seals
    C. worn rear main seal
    D. clogged PCV system

66. The ignition timing is being checked on an older vehicle with a distributor ignition system. To prevent spark advance while the timing is checked, Technician A says that the vacuum hose should be disconnected from the distributor vacuum diaphragm and plugged. Technician B says that, to prevent spark advance while the timing is checked, the timing should be checked when the engine is at idle. Who is correct?
    A. Technician A only
    B. Technician B only
    C. Both A and B
    D. Neither A or B

67. Technician A says that the specific gravity of the battery electrolyte should be checked after adding water to bring all cells to the proper level. Technician B says that specific gravity should be checked right after the battery is charged. Who is correct?
    A. Technician A only
    B. Technician B only
    C. Both A and B
    D. Neither A or B

68. All of the following are true statements regarding a starter current draw test **EXCEPT**:
    A. As a general rule, current draw on a V8 should be higher than on a 4-cylinder engine.
    B. Battery voltage should remain above 9.6V during the test.
    C. Higher than normal current draw readings can be caused by excessive resistance in the starter circuit.
    D. Higher than normal current draw readings can be caused by engine mechanical problems.

69. Technician A says that threaded oil gallery plugs can be removed by first heating them with a torch and then quenching with wax. Technician B says that if a tap-in oil gallery plug comes out, oil pressure will be lost and there will be oil all over the outside of the engine. Who is correct?
    A. Technician A only
    B. Technician B only
    C. Both A and B
    D. Neither A or B

Prepare yourself for ASE testing with these questions on ENGINE REPAIR

70. After reassembling a hydraulic lifter, Technician A says that is should be tested with a leakdown tester. Technician B says that the leakdown rate is how far the lifter plunger moves within a set time period. Who is correct?

    A. Technician A only
    B. Technician B only
    C. Both A and B
    D. Neither A or B

NOTES

Answers to Engine Repair-Test Questions

1. The correct answer is B. The part shown is a simple, flat shim that is placed under the valve spring to adjust its installed height to specification, in response to a change in valve height caused by valve and seat machining.

2. The correct answer is D. If an exhaust valve is not seated, air will leak from the combustion chamber by way of the valve out to the tail pipe and make an audible sound. Broken rings or a bad head gasket would have air leaking through the oil filler or cooling system.

3. The correct answer is C, both technicians. Either method can be used to check main bearing oil clearance.

4. The correct answer is A. The valve stem is hardened and cannot be knurled.

5. The correct answer is A. When the seat contact area is too low on the valve face, 45 degree and 60 degree stones should be used to raise the seat.

6. The correct answer is B. Worn valve guides would cause the gauge needle to fluctuate rapidly at idle but remain steady as the engine speed is increased. A leaking intake manifold gasket or late ignition timing would cause a steady low reading. A weak valve spring would cause the gauge needle to fluctuate as engine speed increased.

7. The correct answer is A. Only technician A is correct. A burned valve will not allow the cylinder to build compression, and the results from wet and dry compression tests will be the same. If piston rings (or worn cylinder walls) were at fault, compression readings from a wet test would exceed dry test results.

8. The correct answer is D, neither technician. Technician A is wrong because incorrect valve timing will cause the valves to open and close at the wrong time, thereby reducing compression. Technician B is wrong because a low but steady manifold vacuum reading may be caused by late ignition timing.

9. The correct answer is C, both technicians. Severe cylinder bore taper can not only be found by measurement, but also by performing a leakage test. After performing a leakage test with the piston at TDC, do a second test with the piston at BDC (Bottom Dead Center). A substantial difference in leakage readings from top to bottom of the cylinder indicates a badly tapered or scored cylinder.

10. The correct answer is B. The distance the valves are opened is defined as valve lift. The time the valves stay open is referred to as duration.

11. The correct answer is C, both technicians. A clearance ramp cushions the take up of clearance between the rocker arm or cam follower. Excessive lash causes the cam lobe to be contacted after the clearance ramp has past. Too little valve lash results in not enough time for heat to be carried from the valve into the seat and its coolant jacket.

12. The correct answer is A. Technician A is correct because most wear to the engine cylinder is at the top where ring loading is the greatest. Technician B is wrong because the difference in bore measurements taken at points parallel and perpendicular to the crankshaft will indicate cylinder bore out-of-roundness.

13. The correct answer is D. Tapping noises are most often caused by loose valvetrain clearances or worn components. However, a tapping sound that may sound like a valvetrain noise can actually be caused by an exhaust leak at the exhaust manifold/cylinder head juncture.

14. The correct answer is A. A bent crankshaft could cause more wear on both bearings in the center of a block; straightening or replacing the crankshaft would be the correct repair. A warped block would cause wear on either the upper or the lower bearing, depending on how the block was warped.

15. The correct answer is C. Worn piston rings will usually make an engine smoke worse under acceleration. All of the other causes can allow oil to be drawn through the valve guides under the high intake vacuum that occurs during deceleration.

16. The correct answer is C, both technicians. Technician A is correct because the cylinders are adjacent to one another, and a blown head gasket between the two cylinders could be causing low compression in both cylinders. Technician B is also correct. Since both cylinders are also consecutive cylinders in the firing order, they may share an ignition problem such as crossed spark plug wires or a carbon track or crack between their terminals inside the distributor cap.

17. The correct answer is C. Installing a new valve seat will position the valve lower in the head and reduce stem height. A and B are incorrect because grinding the valve or seat will place the valve higher in the head and increase the stem height. D is also wrong; shims are used to correct valve spring installed height.

18. The correct answer is B. Magnetic particle detection can only be used on ferrous (iron and steel) parts.

19. The correct answer is C, both technicians. The oil pump should be filled with oil during engine reassembly to prevent it from cavitating or sucking air during engine start-up. To prevent the pickup from sucking air and subsequent oil starvation and engine damage, make sure the pickup screen is positioned close to the bottom of the oil pan to ensure it is always submerged in oil.

20. The correct answer is D. The crankshaft main and rod journals should be inspected and measured for diameter, out-of-round and taper wear.

21. The correct answer is B. Worn pistons can cause noise and accelerate ring and bore wear, but would not cause low oil pressure. Diluted oil, excessive bearing clearance and a clogged pump pickup can all cause low oil pressure.

22. The correct answer is D, neither technician. If a failed exhaust valve was the cause, air would be heard from the exhaust pipe. If the intake valve failed, air would be heard from the carburetor or throttle valve assembly. If bubbles are seen in the radiator coolant, the cause could be a leaking cylinder head gasket or a cracked head or block.

23. The correct answer is D, neither technician. Umbrella seals are installed by pushing the seal down on the valve stem until it touches the valve guide boss. It will be positioned correctly the first time the valve opens. Positive seals are tapped into place on the guide boss using a special installation tool and a small mallet.

24. The correct answer is A. The stem tip height increases with valve or seat grinding. Grinding stem tips can change the stem tip height, but will have no effect on the valve spring installed height. That measurement is controlled by the height of the keeper grooves above the spring seat of the head. Valve spring installed height is corrected by installing shim(s) between the valve spring and cylinder head.

25. The correct answer is C, both technicians. Both will affect valvetrain geometry.

26. The correct answer is B. 'X' and 'Y' are the thrust surfaces of the crankshaft thrust bearing. The clearance between the thrust bearing and crankshaft is called crankshaft end-play and is measured using a dial indicator by pushing the crank fore-and-aft as far as it will go. Crankshaft end-play can also be measured using a feeler gauge.

27. The correct answer is A. Piston rings are designed with a slight taper on the ring face to scrape the oil off the cylinder wall. When installed upside down, rather than scraping oil down and off the walls, they scrape oil up into the combustion chamber where it is burned, causing increased oil consumption. Installing the rings upside down does not change the ring land clearance, so the ring lands will not be affected.

28. The correct answer is C. Manufacturer's specify that piston diameter be measured at a fixed distance below the ring lands, perpendicular to the piston pin. The manufacturer's specifications should always be consulted for specific measuring location.

Answers to Engine Repair-Test Questions

29. The correct answer is A. If not within specification, spring installed height can be corrected by installing shim(s) between the valve spring and cylinder head. If spring free height, squareness or pressure is not within specification, the spring must be replaced.

30. The correct answer is C, both technicians. The lifter face is convex, the cam lobe is slightly tapered (about 0.0005-0.0007-in.) and the lifter bore is offset to make the lifter rotate during operation. This rotation is essential because it spreads wear over a greater surface area. If the lifter fails to spin, it will destroy itself and its cam lobe.

31. The correct answer is A. To adjust valve lash for one cylinder's cam lobes, rotate the crankshaft so that its piston is at TDC on its compression stroke. This will position the intake and exhaust valve lifters or OHC cam followers on the base circle of their respective cam lobes. Hydraulic lifters should be adjusted until zero lash is reached, then turned an additional number of turns (3/4 to 1 1/2 turns is typical). The intent is to position the lifter plunger midway in its travel in the lifter body.

32. The correct answer is D, neither technician. The thermostat's function is to allow the engine to come to operating temperature quickly and then maintain a minimum operating temperature. The PCM bases fuel mixture, ignition timing, EGR operation and other functions in part on input regarding coolant temperature. Only the correct heat range thermostat should be installed. If the thermostat is good and the engine runs hot, another problem exists.

33. The correct answer is C. Both technicians are correct.

34. The correct answer is D, neither technician. Some engines use torque-to-yield head bolts. These bolts are purposely overstretched when tightened and are not reused. Head bolts should be tightened to their full torque value in several incremental steps, rather than all at once.

35. The correct answer is D, neither technician. A new timing chain should always be installed on new sprockets. Unless they are damaged, it is usually not necessary to replace the sprockets when replacing a timing belt.

36. The correct answer is C. Caustic will damage aluminum.

37. The correct answer is B. Magnafluxing is used to detect cracks, not repair them.

38. The correct answer is D. All of the spark plugs should be removed so the engine will crank more easily.

39. The correct answer is C, both technicians. One rod bearing showing wear that is different than the rest could have an obstruction in the crankshaft oil passage(s) that supplies the bearing or an out-of-round rod bore.

40. The correct answer is B. Ignition timing that is too far advanced would cause pinging or detonation, but retarded ignition timing would not. Excessive cylinder head resurfacing would raise compression and could cause pinging or detonation. An inoperative EGR valve would let combustion chamber temperatures become too high and could cause pinging or detonation. Fuel with too low an octane rating will burn too quickly and can be ignited by hot spots in the combustion chamber.

41. The correct answer is C. Both technicians are correct.

42. The correct answer is C, both technicians. The outer ring on a defective harmonic balancer can slip, causing inaccurate ignition timing readings. A broken crankshaft can also result from an incorrect or defective balancer.

43. The correct answer is A. On externally balanced engines, the flywheel must be correctly indexed when it is mounted on the crankshaft to maintain engine balance (many flywheels have bolt holes that are unevenly spaced so they can only go on one way). Make certain that index marks are properly aligned when installing the flywheel. Runout on the face of the flywheel is checked with a dial indicator.

44. The correct answer is D. A thermostat that is stuck open will not restrict coolant flow. A clogged cooling system would restrict coolant flow and cause engine overheating. A defective fan clutch and broken fan shroud would both reduce airflow through the radiator and could cause engine overheating.

45. The correct answer is B. A cracked valve cover gasket and worn rear main seal would allow oil to leak past their damaged surfaces. Worn piston rings would cause excessive crankcase pressure and push oil past seals and gaskets that are in good condition. A stuck open PCV would not cause excessive crankcase pressure, but would cause a vacuum leak and cause the engine to run rough at idle.

46. The correct answer is B. The weights and springs in a centrifugal advance mechanism are calibrated to advance ignition timing as engine speed increases, therefore the engine must be accelerated when checking centrifugal advance. Vacuum advance works only in response to changes in vacuum applied to the unit, therefore changes in ignition timing can be seen with a timing light while applying vacuum with the engine at idle.

47. The correct answer is C, both technicians. A new ring can contact the old rounded ring ridge. This will push the bottom part of the ring groove down, possibly affecting second ring clearance. If the block is to be rebored, removal of the ring ridge will not be necessary.

48. The correct answer is D. A dirty air filter and restricted exhaust would inhibit airflow through the turbocharger and reduce performance. Worn turbocharger shaft bearings could cause the rotating assembly to bind or drag and reduce performance. A stuck open wastegate would reduce performance, however a stuck closed wastegate would cause overboost.

49. The correct answer is C. During a load test, a battery should be loaded to one-half its CCA rating for 15 seconds.

50. The correct answer is A. A sleeve can be used to repair a damaged or cracked cylinder bore.

51. The correct answer is B. It makes no sense to perform repairs on a cylinder head that might not be useable. The head should first be checked for warpage and cracks. If the top of the head is warped enough to interfere with cam bore alignment and/or restrict free movement of the camshaft, the head must be straightened before it is resurfaced.

52. The correct answer is B. Connecting rod journals typically experience the most wear at TDC. Because bearing journals can experience taper wear, barrel wear or hourglass wear, they should be measured at the center and at both ends.

53. The correct answer is C. Resurfacing the block deck will decrease the deck height (the distance from the top of the pistons at TDC to the top of the block) and raise compression.

54. The correct answer is A. A tight belt will cause wear to the front upper main bearing. When an oil supply problem has resulted in a knock, the bearings furthest from the oil pump suffer first.

55. The correct answer is B. Gasket sealer must not be used on synthetic rubber gaskets. It acts like a lubricant, causing the gasket to slip and leak.

56. The correct answer is D, neither technician. When replacing a thermostat, the heat sensing element should be installed in the block or exposed to the hot water side. The fan clutch should be replaced if there is no viscous drag when the fan is rotated.

57. The correct answer is C. Cylinder walls are best cleaned with hot water and a stiff bristle brush. Solvents drive grit deeper into the pores of the cast iron cylinder wall.

58. The correct answer is B. When a cylinder is bored, 0.0025-0.003-in. (0.064mm-0.076mm) cylinder wall material is left for finish honing to final size. This is necessary to completely remove the rough finish left by the boring tool.

Answers to Engine Repair-Test Questions

59. The correct answer is D, neither technician. On engines with full-floating wrist pins, the retainer clips at the ends of the pin should always be replaced with new ones. Full floating pin bushings that are worn are replaced by pressing them in. Then they are finish honed to provide the correct clearance.

60. The correct answer is C. Rings are installed on the piston using a ring expander; they are installed in the cylinder with the piston using a ring compressor.

61. The correct answer is D. If the shop was located at sea level, any of the other answers might be indicated by this vacuum reading. But remember, atmospheric pressure changes with elevation and vacuum readings must be adjusted accordingly. As an approximation for every 1000 ft. above sea level, remove one inch of vacuum. If we correct for elevation, we can see that this engine would produce 19 in. Hg vacuum at idle at sea level, which is within the parameters for normal engine operation.

62. The correct answer is A. When you eliminate the ignition or injection to a cylinder with a rod knock, the sound diminishes. Unlike a connecting rod bearing noise, piston slap does not quiet down and may in fact grow louder when you eliminate ignition or fuel injection to that cylinder.

63. The correct answer is D. A properly operating fan clutch will change the speed of the fan according to engine temperature. When the engine is cold, the fan clutch does not turn the fan very fast, even when engine speed is increased. As the engine warms up, the fan clutch increases the speed of the fan. If the fan speed varies only according to engine speed, regardless of temperature, then the fan clutch is probably seized and should be replaced.

64. The correct answer is C. When performing a leakdown test, if pumping air into a particular cylinder produces bubbles in the radiator coolant, you know that the leak is in that cylinder. The other methods can be used to determine if the engine has an internal coolant leak, but would not identify the leaking cylinder.

65. The correct answer is B. A rear main seal can leak due to wear or age. Worn piston rings can allow combustion blow-by gases to enter the crankcase, where the gases can push oil past seals and gaskets that are in good condition. This type of leak can also be caused by excessive crankcase pressure due to a clogged PCV system. Worn valve seals can cause an engine to smoke, but would not cause an oil leak.

66. The correct answer is C, both technicians. Disconnecting the vacuum hose from the distributor vacuum diaphragm will prevent vacuum advance. At low engine speeds, the springs hold the weights of the centrifugal advance mechanism inward, preventing centrifugal advance. Ignition timing is usually checked at idle or at a specific engine speed before centrifugal advance occurs.

67. The correct answer is D, neither technician. There are times when specific gravity readings will be inaccurate. One is just after adding water to the battery. After adding water, it is recommended to wait at least one day, or until the vehicle has been operated for a while to check specific gravity. Another time is during or just after charging. After charging, it is recommended to wait at least 15 minutes prior to checking specific gravity.

68. The correct answer is C. Amperage that is too low may be caused by starter circuit resistance. All of the other answers are true statements regarding a starter current draw test.

69. The correct answer is A. Technician B is wrong because tap-in oil gallery plugs are usually used inside the engine. If one came out, it would cause a loss of oil pressure but the oil leak would be inside the engine. Oil gallery plugs on the outside of the engine are most often threaded.

70. The correct answer is A. Technician B is wrong because the leak-down rate is the time it takes the plunger to move a certain distance, not the distance the plunger moves within a certain period of time.

NOTES

Glossary of Terms

--a--

abrasion - rubbing away or wearing of a part.

abrasive cleaning - cleaning that requires physical abrasion (e.g., glass bead blasting, wire brushing).

acidity - in lubrication, the presence of acid-type chemicals, which are identified by the acid number. Acidity within oil causes corrosion, sludge and varnish to increase.

actuator - a control device that delivers mechanical action in response to a vacuum or electrical signal; anything that the engine control computer uses to do something, such as trigger fuel injection or fire a spark plug. Most actuators on a computer-controlled engine system are activated by grounding their circuits rather than by actively powering them, since that protects the computer from short circuits.

additive - material added to the engine oil to give it certain properties.

adhesion - an oil's ability to cling to a surface.

aeration - the process of mixing air with liquid.

aerobic - curing when exposed to oxygen.

air duct - a tube, channel or other tubular structure used to carry air to a specific location.

air gap - space or gap between spark plug electrodes, motor and generator armatures and field shoes.

air injection reaction (AIR) system - a system that provides fresh air to the exhaust system under controlled conditions to reduce emissions. The air source can be a pulse-air pump or an electrically or belt driven pump. Upstream air injection goes into the exhaust manifold to assist in after-burning HC laden exhaust gases. Downstream air injection goes into the oxidation bed of the catalytic converter to help oxidize HC and CO emissions.

air lock - a bubble of air trapped in a fluid circuit that interferes with normal circulation of the fluid.

air pump - device to produce a flow of air at higher-than-atmospheric pressure. Normally referred to as a thermactor air supply pump.

align boring - a machining method that realigns bearing bores to center and makes the bores round.

alignment - an adjustment to bring into a line.

alloy - a mixture of different metals (e.g., solder, an alloy consisting of lead and tin).

aluminum - a light weight metal used often for cylinder heads and other parts.

ambient temperature - the temperature of the air surrounding an object.

ammeter - instrument used to measure electrical current flow in a circuit.

ampere (amp) - unit for measuring electrical current.

anaerobic - curing in the absence of oxygen.

antifreeze - a material such as ethylene glycol which is added to water to lower its freezing point; used in an automobile's cooling system.

arcing - electrical energy jumping across a gap.

atmospheric pressure - the weight of the air at sea level (14.7 lbs. per sq. in.) or at higher altitudes.

axial - having the same direction or being parallel to the axis of rotation.

axial load - a type of load placed on a bearing that is parallel to the axis of the rotating shaft.

axial play - movement parallel to the axis of rotation.

--b--

babbitt - a soft bearing alloy used in engine bearings.

backlash - the clearance or play between two parts, as in gear mesh.

back pressure - pressure created by restriction in an exhaust system.

balance shaft - weighted shaft used on some engines to reduce vibration.

ball bearing - an anti-friction bearing that uses a series of steel balls held between inner and outer bearing races.

barometric pressure - the pressure of the atmosphere, usually expressed in terms of the height of a column of mercury. A sensor or its signal circuit sends a varying frequency signal to the processor relating actual barometric pressure.

base circle - the part of a camshaft lobe that is opposite the tip of the nose. The part of the camshaft lobe that does not move the valve in any way.

battery - a device that produces electricity through electrochemical action.

battery acid - the sulfuric acid solution used as the electrolyte in a battery.

battery cell - the part of a storage battery made from two dissimilar metals and an acid solution. A cell stores chemical energy to be used later as electrical energy.

bearing - part that supports and reduces friction between a stationary and moving part or two moving parts.

bearing cap - the lower half of the bearing saddle. It is removable, as in main cap or rod cap.

bearing clearance - the space between a bearing and its corresponding component's loaded surface. Bearing clearances are commonly provided to allow lubrication between the parts.

bearing crush - the bearing is slightly larger at its parting lines so that when the two halves of the bearing are tightened together, the bearing seats firmly in its bore. This keeps the bearing from rotating.

bearing lining - the bearing surface area of a bearing, it is usually made up of an alloy of several metals, including lead.

bearing race - machined circular surface of a bearing against which the roller or ball bearings ride.

bearing spread - the condition in which the distance across the outside paring edges of the bearing insert is slightly greater than the diameter of the housing bore.

before top dead center (BTDC) - the degrees of crankshaft rotation just before the piston in a specific cylinder reaches TDC, the highest point in its vertical travel on the compression stroke. On most vehicles, spark occurs a certain number of degrees of crankshaft rotation BTDC.

big end - the large end of the connecting rod that connects to the crankshaft.

blowby - the unburned fuel and products of combustion that leak past the piston rings and into the crankcase at the last part of the combustion stroke.

bolt diameter - measurement across the major diameter of a bolt's threaded area or the bolt shank.

bolt head - part of the bolt that the socket or wrench fits over in order to torque or tighten the bolt.

boost pressure - term used when a turbocharger increases the air pressure entering an engine above atmospheric pressure.

bore - a cylindrical hole.

bore gauge - a precision measuring instrument used to measure the diameter of a bore.

boss - the part of a piston that fits around its pin.

bottom dead center (BDC) - when the piston is at its lower limit in the cylinder bore.

bracket - used to secure parts to the body or frame.

break-in - a slow wearing-in process between two mating part surfaces.

--C--

camshaft - a shaft with eccentric lobes that control the opening of the intake and exhaust valves.

camshaft bearing - a bearing that supports the camshaft journal. On some engines it is full round and pressed in place. On some OHC engines the camshaft bearing is made up of two shells like a connecting rod bearing.

camshaft follower - on OHC engines the equivalent of a rocker arm.

Glossary of Terms

camshaft journal - the bearing area of a camshaft.

camshaft lobe - the eccentric on a camshaft that acts on lifters or followers and in turn, other valvetrain components as the camshaft is rotated, to open the intake and exhaust valves.

camshaft sprocket - the sprocket on a camshaft that is turned by a chain or belt from the crankshaft. The camshaft sprocket has twice as many teeth as the crankshaft sprocket.

carbon - a hard or soft nonmetallic element that forms in an engineís combustion chamber when oil is burned.

carbon dioxide (CO_2) - a colorless, odorless, noncombustible gas, heavier than air; can be compressed into a super-cold solid known as dry ice; changes from solid to vapor at −78.5°C.

carbon monoxide (CO) - a colorless, odorless gas, which is highly poisonous. CO is produced by incomplete combustion. It is absorbed by the bloodstream 400 times faster than oxygen.

carbonize - the process of carbon formation within an engine, such as on the spark plugs and in the combustion chamber.

carburetor - a device that atomizes air and fuel in a proportion that is burnable in the engine.

casting - metal that is manufactured by pouring it into a mold. It is more porous than forged metal and does not conduct heat as well. It is less expensive to manufacture, however.

catalyst - a compound or substance that can speed up or slow down the reaction of other substances without being consumed itself. In an automatic catalytic converter, special metals (platinum or palladium) are used to promote combustion of unburned hydrocarbons and reduce carbon monoxide.

catalytic converter - an emission control device located in the exhaust system that contains catalysts, which reduce hydrocarbons, carbon monoxide and nitrogen oxides in the exhaust gases.

centrifugal force - the force which pulls an object outward when it is rotating rapidly around a center.

chafing - damage caused by friction and rubbing.

chamfer - a bevel or taper at the edge of a hole or corner, usually cut at 45 degrees.

chase - to straighten or repair damaged threads.

check valve - a gate or valve that allows passage of gas or fluid in one direction only.

chemical cleaning - relies primarily on chemical action to remove dirt, grease, scale, paint or rust.

clearance - the specified distance between two components.

closed loop - electronic feedback system in which sensors provide constant information on what is taking place in the engine.

coefficient of friction - a relative measurement of the friction developed between two objects in contact with each other.

cold cranking amps (CCA) - the amount of cranking amperes that a battery can deliver in 30 seconds at 0°F (−18°C).

combustion - the burning of the air/fuel mixture.

combustion chamber - enclosure formed by a pocket in the cylinder head and the top of the piston, where the spark plug ignites the compressed air/fuel mixture. The volume of the cylinder above the piston when the piston is at TDC.

comparator gauge - a metal card with sample patches of various surface textures to visually compare with a milled or ground surface.

compression - in a solid material, compression is the opposite of tension. In a gas, compression causes the gas to be confined in a smaller area, raising its temperature and pressure.

compression rings - usually the top two rings on a piston, they form a seal between the piston and cylinder wall to compress the air fuel mixture in the cylinder.

connecting rod - a rod that connects the crankshaft to the piston and enables the reciprocating motion of the piston to turn the crankshaft.

connecting rod bearings - the plain bearing shells located in the big end of the connecting rod that support the connecting rod and piston on the crankshaft.

connecting rod cap - the removable part of the big end of the connecting rod.

coolant - mixture of water and ethylene glycol-based antifreeze that circulates through the engine to help maintain proper temperatures.

cooling fan - a mechanically or electrically driven propeller that draws air through the radiator.

cooling system - the system used to remove excess heat from an engine and transfer it to the atmosphere. Includes the radiator, cooling fan, hoses, water pump, thermostat and engine coolant passages.

compound - a mixture of two or more ingredients.

compression ratio - ratio of the volume in the cylinder above the piston when the piston is at bottom dead center to the volume in the cylinder above the piston when the piston is at top dead center.

compression stroke - the second stroke of the 4-stroke engine cycle, in which the piston moves from bottom dead center and the intake valve is closed, trapping and compressing the air/fuel mixture in the cylinder.

concentric - two or more circles which have a common center.

contraction - reduction in mass or dimension; the opposite of expansion.

core - in automotive terminology, the main part of a heat exchanger, such as a radiator, evaporator or heater. Usually made of tubes, surrounded by cooling fins, used to transfer heat from the coolant to the air.

core plugs - plugs that fill holes in a block or head left from the casting process. Also called freeze, welsh or expansion plugs.

corrode - gradual loss from a metal surface from chemical action.

corrosion - the eating into or wearing away of a substance gradually by rusting or chemical action.

corrosivity - the characteristic of a material that enables it to dissolve metals and other materials or burn the skin.

counterbore - to enlarge a hole to a given depth.

countersink - to cut or form a depression to allow the head of a screw to go below the surface.

counterweight - weights that are part of a crankshaft casting or forging. They counterbalance the weight of the connecting rods and journals to reduce vibration.

crankcase - the lower part of an engine block that houses the crankshaft.

crankpin - machined, offset area of a crankshaft where the connecting rod journals are machined.

crankshaft - a lower engine part with main and rod bearing journals. It converts reciprocating motion to rotary motion.

crankshaft journals - the bearing areas of a crankshaft are the main and rod journals.

crankshaft pulley - the belt drive pulley mounted on the front of the vibration damper.

crankshaft thrust collar - a flat machined area that is 90 degrees to the crankshaft main journals. The flange of the thrust main bearing rides against it to control crankshaft end thrust.

crank throw - distance from the crankshaft main bearing centerline to the connecting rod journal centerline. The stroke of any engine is the crank throw.

crank web - unmachined portion of a crankshaft that lies between two crank pins or between a crankpin and main bearing journal.

crosshatch - the pattern left on a finished piston wall, retains oil and aids piston ring seating.

crush - the bearing is slightly larger at its parting lines so that when the two halves of the bearing are tightened together, the bearing seats firmly in its bore. This keeps the bearing from rotating.

cylinder - a round hole in the engine block for the piston.

cylinder balance test - an engine diagnostic test used to compare the power output of all the engine's cylinders. Also known as a power contribution test.

cylinder bore - a cylindrical hole.

Glossary of Terms

cylinder head - the casting that contains the valves and valve springs, and covers the top of the cylinders.

cylinder leakage test - an engine diagnostic test where the piston in the cylinder to be tested is brought to top dead center (TDC) on the compression stroke and compressed air is pumped into the cylinder through the spark plug hole. Where the air leaks out shows the location of the compression leak. A leakage tester will compare the air leaking out of the cylinder to the amount of air being put into it, expressed as a percentage.

cylinder sleeve - a replacement iron liner that fits into a cylinder bore. It can be either wet or dry. The outside diameter of wet sleeves contacts the coolant.

cylinder walls - the walls of the cylinder bore.

current - the number of electrons flowing past a given point in a given amount of time.

--d--

dampen - to slow or reduce oscillations or movement.

damper - a device mounted on the front of the crankshaft. It reduces the torsional or twisting vibration that occurs along the length of the crankshaft in multiple cylinder engines. It is also called a harmonic balancer.

dead center - the extreme upper or lower position of the crankshaft throw at which the piston is not moving in either direction.

deck - top of the engine block where the cylinder head is mounted.

deflection - bending or movement away from the normal position due to loading.

degree - used to designate temperature readings or 1 degree as a 1/360 part of a circle.

density - relative mass of matter in a given volume.

detergent - a compound of soap-like nature used in engine oil to remove engine deposits and hold them in suspension in the oil.

detonation - abnormal combustion of an air fuel mixture. When pressure in the cylinder becomes excessive and the mixture explodes violently, instead of burning in a controlled manner. The sound of detonation can be heard as the cylinder walls vibrate. Detonation is sometimes confused with preignition or ping.

dial caliper - versatile instrument capable of taking inside, outside, depth and step measurements.

dial indicator - a measuring device equipped with a readout dial used most often to determine end motion or irregularities.

diaphragm - flexible, impermeable membrane on which pressure acts to produce mechanical movement.

direct ignition - a distributorless ignition system in which spark distribution is controlled by the vehicle's computer.

displacement - the volume a cylinder holds between top dead center and bottom dead center positions of the piston.

dowel - pin extending from one part to fit into a hole in an attached part; used for location and retention.

driveability - the degree to which a vehicle operates properly, including starting, running smoothly, accelerating and delivering reasonable fuel mileage.

dry sleeve - a sleeve, that when installed in a cylinder block, does not come into contact with coolant.

duration - the length of time that a valve remains open, measured in crankshaft degrees.

dwell time - degree of crankshaft rotation during which the primary circuit is on.

dykem blue - a dye which is painted on a valve seat in order to determine seat concentricity. The valve is inserted into the guide, lightly seated, and rotated about 1/8-in. (3.175mm). A continuous blue line should appear all the way around the valve face if the valve and seat are mating properly. Open patches or breaks in the line indicate that the seat is not concentric and the low spots are not making contact.

--e--

eccentric - the part of a camshaft that operates the fuel pump.

efficiency - ratio of the amount of energy put into an engine compared to the amount of energy coming out of the engine; a measure of how well a particular machine works.

elasticity - the principle by which a bolt can be stretched a certain amount. Each time the stretching load is reduced, the bolt returns to exactly its original, normal size.

electrode - firing terminals found in a spark plug.

electrolysis - chemical and electrical decomposition process that can damage metals such as brass, copper and aluminum in the cooling system.

electrolyte - a material whose atoms become ionized (electrically charged) in solution. In automobiles, the battery electrolyte is a mixture of sulfuric acid and water.

electromagnet - a magnet formed by electrical flow through a conductor.

electromagnetic induction - moving a wire through a magnetic field to create current flow in the wire.

electromechanical - refers to a device that incorporates both electrical and mechanical principles together in its operation.

electronic - pertaining to the control of systems or devices by the use of small electrical signals and various semiconductor devices and circuits.

embedability - the ability of the bearing lining material to absorb dirt.

emulsion - mixture of air and fuel in the carburetor.

end-play - the amount of axial or end-to-end movement in a shaft due to clearance in the bearings.

energy - the ability to do work.

engine - a device that converts heat energy into mechanical energy.

engine block - the casting made up of cylinders and the crankcase.

engine displacement - the volume displaced by all of the pistons in all of an engine's cylinders.

excessive wear - wear caused by overloading a part that is in an out-of-balance condition, resulting in lower-than-normal life expectancy of the part being subjected to the adverse operating condition.

exhaust gas recirculation (EGR) system - helps prevent the formation of oxides of nitrogen (NOx) by recirculating a certain amount of exhaust as an inert gas through the intake manifold to keep the peak combustion temperatures below what would form those chemical compounds. The computer determines when and how much exhaust to recirculate based on information from all its other sensors. It then actuates the EGR solenoid, which opens a vacuum circuit or operates an electronic circuit or operates an electronic circuit to actually work the EGR valve. The computer uses a duty-cycle (percentage of on-time) signal to activate the solenoid.

exhaust gas recirculation (EGR) valve - component in the EGR system, used to meter a controlled amount of exhaust gas into the intake air stream.

exhaust manifold - the part of the exhaust system that is fastened to the cylinder head.

exhaust pipe - the pipe between the exhaust manifold and muffler.

exhaust port - the passage or opening in a 4-stroke cylinder head for the exhaust valve.

exhaust stroke - the final stroke of the 4-stroke engine cycle, in which the compressed fuel mixture is ignited in the combustion chamber.

expansion - an increase in size.

--f--

fan - a mechanically or electrically driven propeller that draws or pushes air through the radiator, condenser, heater core or evaporator core.

fan clutch - a device attached to a mechanically driven cooling fan that allows the fan to freewheel when the engine is cold or the vehicle is driven at speed.

fan shroud - an enclosure that routes air through the radiator cooling fins.

Glossary of Terms

fatigue - deterioration of a bearing metal under excessive intermittent loads or prolonged operation; in mechanical terms, the tendency of a material, especially metal, to fail under repeated applications of stress.

fatigue strength - a bearing's ability to withstand loads.

feeler gauge - thin metal strip manufactured in precise thickness and used to measure clearance between parts; usually part of a set.

ferrous - containing or derived from iron.

ferrous metal - metal that contains iron or steel and is subject to rust.

flange - a projecting rim or collar on an object for keeping it in place.

flare - expanded, shaped end on a metal tube or pipe.

flexible hone - a hone used primarily to deglaze cylinder walls. Also known as a ball hone.

floating pin - a piston pin that moves in the piston and the small end of a connecting rod. It is held in place by retaining clips in the pin bosses.

flutter - as applied to engine valves, refers to a condition wherein the valve is not held tightly on its seat during the time the cam is not lifting it.

flywheel - a cast iron or steel wheel mounted to the end of the crankshaft; helps to smooth the engine's power delivery, the teeth around its circumference provide an engagement for the starter, and it provides the mounting points for the pressure plate and friction surface for the clutch disc.

foot pound - a unit of measurement for torque. One foot pound is the torque obtained by a force of one pound applied to a wrench handle that is 12-in. long.

force - a pushing effort measured in pounds.

forge - to shape metal; to stamp into a desired shape.

four stroke cycle engine - an engine, either gasoline or diesel that uses four strokes: intake, compression, power and exhaust. A firing impulse occurs every two turns of the crankshaft. When this engine is a gasoline engine, it is also called an Otto cycle engine after its inventor. A diesel engine is called a Diesel cycle engine for the same reason.

free-play - the measurable travel in a mechanical device between the time force is applied and work is accomplished; looseness in a linkage between the start of application and the actual movement of the device, such as the movement in the steering wheel before the wheels start to turn.

freewheel - a mechanical device that engages the driving member to impart motion to a driven member in one direction but not the other.

freeze plug - another name for core plug.

friction - resistance to motion that occurs when two objects rub against each other.

fuel pressure regulator - uses intake manifold vacuum, or more properly intake manifold absolute pressure (MAP), to modify the pressure in the fuel rail. The fuel pump can pump more fuel than the engine can use, so the system routes the extra fuel back to the tank through the fuel pressure regulator.

fulcrum - the support or point on which a lever rests; also called 'pivot point'.

--g--

gasket - a material such as artificial rubber, cork, or steel used to seal between parts that would otherwise leak fuel, coolant, lubricants or combustion gases.

gear - a toothed wheel, disc, etc. designed to mesh with another or with the thread of a worm; used to transfer or change motion.

gear pitch - number of teeth per given unit of pitch diameter. Gear pitch is determined by dividing the number of teeth by the pitch diameter of the gear.

gear pump - positive displacement pump that uses two meshing external gears, one drive and one driven.

glaze - thin residue on cylinder walls formed by a combination of heat, engine oil and piston movement.

glass bead blasting - an operator intensive abrasive cleaning method that uses glass beads propelled by compressed air in an enclosed blasting cabinet.

glaze breaker - a spring loaded tool with honing stones that restores the crosshatch surface of a cylinder wall prior to installing new piston rings.

go/no-go gauge - a measuring tool that tells if a tolerance is met or not.

grind - to machine a surface with an abrasive wheel.

ground - negatively charged side of a circuit; can be a wire, negative side of the battery or vehicle chassis.

--h--

harmonic balancer - a device that reduces the torsional or twisting vibration that occurs along the length of the crankshaft in multiple cylinder engines. It is also called a vibration damper.

harmonic vibration - periodic motion or vibration along a straight line. The severity depends on the frequency or amplitude.

harmonics - potentially damaging vibration in the crankshaft or valve springs.

heat dam - narrow groove cut into the top of the piston to restrict the flow of heat into the piston.

heli-coil - one type of thread repair device that consists of a spring loaded stainless steel coil.

hemispherical combustion chamber - a combustion chamber shape that resembles a half a globe. The valves are on the sides of the spark plug, which is in the center.

Hg - the chemical symbol for the element mercury. Engine vacuum is measured in inches of mercury or in/Hg.

high tension - high voltage. In an ignition system, voltages in the secondary circuit of the system as opposed to the low, primary circuit voltage.

hone - abrasive tool for correcting small irregularities or differences in diameter in a cylinder, such as an engine cylinder or brake caliper; to enlarge or smooth a bore with a rotating tool containing an abrasive material.

horsepower (HP) - measurement of an engine's ability to perform work. One horsepower is the energy required to lift 550 pounds 1 foot in 1 second.

housing bore - the machined bore in a block or head where a bearing will be installed.

hydraulic valve lifter - an automatic lash adjusting device that provides a rigid connection between the camshaft and valve, while absorbing the shock of motion. A hydraulic valve lifter differs from the solid type in that it uses oil to absorb the shock that results from movement of the valvetrain.

hydrocarbons (HC) - solid particles of gasoline present in the exhaust and in crankcase vapors that have not been fully burned.

hydrometer - an instrument used to measure the specific gravity of a solution.

--i--

ID - inside diameter.

induction - flow of electrons caused when magnetic lines of force pass across a conductor, as when the coil primary field collapses across the secondary windings.

inertia - the constant moving force applied to carry the crankshaft from one firing stroke to the next.

insert bearing - a bearing made as a self-contained part and then inserted into the bearing housing.

insert guides - valve guides that are pressed fit in the cylinder head.

installed spring height - distance from the valve spring seat to the underside of the retainer when it is assembled with keepers and held in place.

installed stem height - distance from the valve spring seat to the stem tip.

Glossary of Terms

insulated circuit - a circuit that includes all of the high-current cables and connections from the battery to the starter motor.

intake stroke - first stroke of the 4-stroke engine cycle in which the piston moves away from top dead center and the intake valve opens.

intake valve - also called inlet valve, it closes off the intake port and opens it at the correct time in response to movement from the cam lobe.

integral - made of one piece.

integral guides - valve guides that are part of the cylinder head.

integral seats - valve seats that are part of the head.

intercooler - device used on some turbocharged engines to cool the compressed air.

--j--

jet clean - a cleaning machine that sprays engine parts with degreasing solution under high pressure. The parts rotate on a carousel during the cleaning process to expose all surfaces to the cleaning spray.

journal - the bearing surface on a shaft.

--k--

keepers - small locks that hold the valve retainer onto the valve stem. Also called split locks.

keep-alive memory - a series of vehicle battery-powered memory locations in the microcomputer that store information on input failure, identified in normal operations for use in diagnostic routines; adapts some calibration parameters to compensate for changes in the vehicle system.

key - a small block inserted between the shaft and hub to prevent circumferential movement.

keyway - a slot cut into a shaft to accept a key.

knurling - technique used for restoring the inside diameter dimensions of a worn valve guide by plowing tiny furrows through the surface of the metal.

--l--

land - the areas between the grooves of a piston.

lapping - the process of fitting one surface to another by rubbing them together with an abrasive material between the two surfaces.

lash - the amount of clearance between components in a geartrain or valvetrain.

lash adjuster - a device for adjusting valve lash or maintaining zero lash in certain types of OHC engines. The lash adjuster is stationary in the cylinder head, with one end of a cam follower mounted on top of it. The other end of the follower acts on the valve stem when the camshaft lobe, which is positioned over the center of the follower, pushes the follower down.

leakdown - the relative movement of the plunger with respect to the hydraulic valve lifter body after the check valve is seated by pressurized oil. A small amount of oil leakdown is necessary for proper hydraulic valve lifter operation.

lifter - the valvetrain part that rides on the camshaft lobe.

lifter bores - the holes in an engine block that the lifters fit into.

line contact - contact made between the cylinder and the torsional rings, usually on one side of the ring; the contact made between the valve and the valve seat. When an interference angle is used, only a small line of contact is produced.

liner - a thin layer, used as a wear surface or a replaceable guide liner or cylinder sleeve.

load - in mechanics, the amount of work performed by an engine; specifically, the external resistance applied to the engine by the machine it is operating. In electrical terms, the amount of power delivered by a generator, motor, etc., or carried by a circuit. The work an engine must do, under which it operates more slowly and less efficiently (e.g., driving up a hill, pulling extra weight).

lobe - the eccentric part of the camshaft that moves the lifter.

lock stitch - a crack repair method.

lubrication - the process of introducing a friction reducing substance between moving parts to reduce wear.

--m--

magnet - any body that attracts iron or steel.

magnetic field - the areas surrounding the poles of a magnet which are affected by its forces of attraction or repulsion.

magnetic particle detection - a process, often called magnaflux that is used with iron or steel parts to detect cracks.

main bearing caps - the removable lower halves of the main bearing bores.

main bearing clearance - the clearance between the main bearing journal and its bearings.

main bearing journal - the central, load-bearing points along the axis of a crankshaft, where the main bearings support the shaft in the block.

main bearings - the plain bearings that support the crankshaft in the engine block.

main bearing saddle bores - the housings that are machined for main bearings.

major and minor thrust surfaces - the sides of a piston. The major thrust surface receives most of the load from crankshaft rotation.

manifold absolute pressure - measure of the degree of vacuum or pressure within an intake manifold.

manifold absolute pressure (MAP) sensor - a sensor that measures changes in intake manifold pressure resulting from changes in engine load and speed. The pressure in the intake manifold as referenced to a perfect vacuum. Manifold vacuum is the difference between MAP and atmosphere pressure. For example, in a standard atmosphere (sea level) the pressure is 29.92 inches of mercury, 101 kilopascals, or 0 inches of vacuum.

manifold vacuum - relatively low pressure in an engine's intake manifold just below the throttle plate(s). Manifold vacuum is highest at idle and drops during acceleration.

margin - the area between the valve face and the head of the valve.

mass airflow (MAF) sensor - a sensor in a fuel injection system that measures the mass (weight/density) of the incoming air flowing through a meter. The measurement transmitted to the PCM is usually either a frequency or a voltage.

mechanical efficiency - ratio between the indicated horsepower and the brake horsepower of any given engine.

memory - part of a computer that stores or holds programs and other data.

mesh - to fit closely together or interlock, as the fit of gear teeth.

micrometer - a precision measuring instrument. When a micrometer measures in thousandths of an inch, one turn of its thimble results in 0.025-in. movement of its spindle. There are 40 threads per inch ($1/40$th inch = 0.025-in.)

mill - machining with rotating tooth cutters.

millimeter - the base of metric size measurement. One millimeter equals 0.039370-in. One inch is equal to 25.4mm.

misfiring - failure of an explosion to occur in one or more cylinders while the engine is running; can be continuous or intermittent failure.

missing - a lack of power in one or more cylinders.

multimeter - a tool that combines the functions of a voltmeter, ohmmeter and ammeter into one diagnostic instrument.

multiviscosity oil - chemically-modified oil that has been tested for viscosity at cold and hot temperatures.

mushroom lifters - lifters with contact faces that are wider than the lifter bore. They must be installed through the bottom of the lifter bores, before the cam is installed.

mushroomed valve stem tip - a folding over of the metal at the tip of the valve stem in response to pounding from too loose a valve adjustment or a defective hydraulic lifter.

Glossary of Terms

--n--

normal wear - the average expected wear when operating under normal conditions.

normally aspirated - the method by which an internal combustion engine draws air into the combustion chamber. As the piston moves downward in the cylinder, it creates a vacuum that draws air into the combustion chamber through the intake manifold.

--o--

octane - rating indicating a fuel's tendency to resist detonation.

OD - outside diameter.

offset - positioned off center or at an angle; a curve or bend in a metal bar to permit it to pass an obstruction.

OHC engine - overhead cam engine. An engine with the camshaft located in the cylinder head.

ohm - a unit of measured electrical resistance.

oil clearance - the difference between the inside bearing diameter and the journal's diameter.

oil cooler - a device used to remove heat from the engine or transmission oil. There are oil-to-air coolers and oil coolers that are incorporated into the vehicle's cooling system.

oil gallery - a line that supplies oil to areas of the engine block or cylinder head.

oil groove - a groove machined in the bearing surface that provides a channel for oil flow.

oil pan - the part that encloses the crankcase at the lower end of the block.

oil pressure - the pressure that results from resistance to flow from the oil pump. As the pump turns faster, it produces more flow. A relief valve limits the amount of pressure it can produce.

oil pump - the pump that circulates lubricating oil throughout the engine, usually driven by the camshaft (by way of the distributor).

oil pump pickup - the screen that filters and keeps debris out of the oil pump.

oil rings - the bottom ring on the pistons, scrapes excess oil from the cylinder walls to keep it from entering and burning in the combustion chamber.

open loop - an electronic control system in which sensors provide information, the microcomputer gives orders, and the output actuators carry out the orders without feedback to the microcomputer.

O-ring seal - a sealing ring, usually made of rubber and installed in a groove; a type of valve seal that fits into a valve stem groove under the valve keepers.

out-of-round - refers to an inside or outside diameter that was originally designed to be perfectly round, but instead has varying diameters when measured at different points across its diameter.

overbore - the dimension by which a machined hole is larger than the standard size.

overhead cam (OHC) engine - an engine with the camshaft located in the cylinder head.

overhead valve engine - an I-head engine. The intake and exhaust valves are located in the cylinder head.

overlap - the point at TDC where both valves are open at the same time. The intake valve is just beginning to open while the exhaust valve is just finishing closing.

oxidation - the process of combining with oxygen, resulting in rusting or burning. Rust is slow oxidation; fire is rapid oxidation.

oxides of nitrogen (NOx) - various compounds of oxygen and nitrogen that are formed in the cylinders during combustion, and are part of the exhaust gas.

--p--

parting face - the surface of a bearing half that contacts the other bearing half when the bearing is assembled.

parting line - the mark left during casting or forging where the halves of the mold were joined.

PCV valve - a part of the positive crankcase ventilation system. Meters crankcase vapors into the intake manifold.

peen - to stretch or clinch over by pounding with the rounded end of a hammer.

pilot journal - a mechanical journal that slides inside the pilot bearing and guides its movement.

pin boss - the part of a piston that fits around its pin.

pinning - a crack repair method, also called stitching or pegging.

piston - the cylindrical component that is attached to the connecting rod and moves up and down in the cylinder bore. The top of the piston forms the bottom of the combustion chamber. When combustion occurs, the piston is forced downward in the cylinder, moving the connecting rod which in turn rotates the crankshaft.

piston collapse - when the diameter of the piston skirt becomes less due to heat.

piston head - the part of the piston that is above the rings.

piston pin - (see *wrist pin*).

piston ring - an open-ended ring which fits into a groove on the outer diameter of the piston. Its chief function is to form a seal between the piston and cylinder wall. Most automotive pistons have three rings: two for compression sealing; one for oil sealing.

piston slap - a noise that result from excessive piston to cylinder wall clearance.

pitch - in machinery, the distance between corresponding points on two adjacent gear teeth, or threads of a screw or bolt measured along the axis; the angle of the valve spring twist. A variable pitch valve spring has unevenly-spaced coils.

pitting - surface irregularities caused by corrosion.

plain bearing - a type of bearing where the load is supported on a thin film of pressurized oil.

Plastigage™ - a plastic material that is compressed between a bearing and journal, and the resulting compressed material measured to determine the clearance.

play - movement between two parts.

polarity - the particular state (positive or negative) with reference to the two magnetic poles.

poppet valve - a valve consisting of a round head with a tapered face, an elongated stem that guides the valve, and a machined slot at the top of the stem for the valve spring retainer.

porosity - tiny holes in casting caused by air bubbles.

ports - valve openings in a cylinder head.

positive crankcase ventilation (PCV) system - a system that controls crankcase emissions by using a valve to meter crankcase vapors into the intake manifold.

positive displacement pump - oil pump through which a fixed volume of oil passes with each revolution of its driveshaft.

positive seal - a type of valve seal that fits tightly around the top of the valve guide.

power - a measure of work being done.

power stroke - the third stroke of the 4-stroke engine cycle, in which the compressed fuel mixture is ignited in the combustion chamber (see *intake stroke* and *exhaust stroke*).

powertrain control module (PCM) - on vehicles with computer control systems, the main computer that determines engine operation based on sensor inputs and by using its actuator outputs. The PCM may also control transmission operation.

preignition - also called ping, it is abnormal combustion of the air/fuel mixture before it is time to do so. A hot surface or carbon deposit in the combustion chamber ignites the air/fuel mixture before the spark plug is fired. The sound of preignition can be heard as the cylinder walls vibrate. Detonation is sometimes confused with preignition.

preload - tightening a bearing a specified amount past zero lash to eliminate axial play.

Glossary of Terms

press fit - when a part is slightly larger than a hole it must be forced together with a press.

pressure - the exertion of force upon a body, measured in pounds per square inch on a gauge.

primary circuit - the low-voltage circuit of an ignition system.

profilometer - an instrument used to measure surface profiles and surface roughness.

Prussian blue - a paste used to determine the contact area between two parts, such as the height of the valve seat on the valve face.

pushrod - a rod between the lifter and rocker arm. They are sometimes hollow to allow oil distribution to the valves.

--r--

race - channel in the inner or outer ring of an anti-friction bearing in which the balls or rollers operate.

radial - perpendicular to the shaft or bearing bore.

radial load - load applied at 90 degrees to an axis of rotation.

radiator - the part of the cooling system that acts as a heat exchanger, transferring heat to atmosphere. It consists of a core and holding tanks connected to the cooling system by hoses.

radiator cap - a device that seals the radiator and maintains a set pressure in the cooling system.

ratio - proportion of one number to another.

reaming - technique used to repair worn valve guides by increasing the guide hole size to take an oversized valve stem, or by restoring the guide to its original diameter.

rear main oil seal - a seal that fits around the rear of the crankshaft to prevent oil leaks.

reciprocating - up-and-down or back-and-forth motion.

residue - surplus; what remains after a separation.

resistance - the opposition offered by a substance or body to the passage of electric current through it.

ridge - a raised area at the top of a cylinder bore created by ring wear. The ridge occurs because the piston ring does not travel all the way to the top of the bore, thereby leaving an unused portion of cylinder bore above the limit of ring travel. This ridge will usually be more pronounced on high mileage engines.

ridge reamer - a tool used to remove the ridge from the top of a cylinder bore.

rigid hone - a hone that removes metal and imparts a precise finish and crosshatch to the bore.

ring end gap - the clearance between the ends of a piston ring when installed in the cylinder bore.

ring file - a tool used to trim the ends of a piston ring to bring the ring end gap within specification.

ring land - high parts between grooves of a piston.

ring lands - the raised parts between the piston ring grooves.

rocker arm - a pivot lever mounted on a round shaft or a stud. One end of the rocker arm is applied by the pushrod and the other end acts upon the valve stem.

rocker shaft - a round pipe that is mounted parallel on top of the cylinder head. All of the rocker arms on the head are mounted on it.

rocker stud - a stud that is pressed or threaded into a cylinder head on which the rocker arm is mounted.

roller bearing - an anti-friction device made up of hardened inner and outer races between which steel rollers move.

roller lifter - lifters that are equipped with rollers at the bottom that ride on the camshaft lobe, in order to reduce friction.

rotary - refers to a circular motion.

rotor pump - type of oil pump that utilizes a 4-lobe inner rotor and a 5-lobe outer rotor. Output per revolution depends upon rotor diameter and thickness.

runner - a cast tube on an intake or exhaust manifold used to carry air in or out of the engine.

runout - degree of wobble outside normal plane of rotation.

--s--

score - a scratch, ridge or groove marring a finished surface.

scuffing - scraping and heavy wear from the action of a piston on cylinder walls.

seal - a part, usually made of rubber or plastic, installed around a moving part or shaft to prevent leaks.

seat - a surface (usually machined) upon which another part rests or seats (e.g., valve seat).

secondary circuit - the high voltage side of the ignition system, usually above 20,000 volts. The secondary circuit includes the ignition coil, coil wire, distributor cap, rotor, spark plug wires and spark plugs.

seize - when a part sticks, preventing the engine from turning. An example is when a piston welds itself to a cylinder wall because of insufficient clearance or lubrication.

serpentine belt - a flat, ribbed drive belt that makes multiple angles, driving several components.

shim - thin sheets of a material, such as metal, which are used as spacers between two parts.

siamese cylinders - when two cylinders are joined at one side without a coolant jacket between them.

siamese ports - when two cylinders are fed through one port.

skirt - the sides of the piston that are against the cylinder walls.

sleeve - a thin metal liner, such as is commonly used in a cylinder bore; the outer part of the synchronizer assembly. When shifting gears, the sleeve moves along the splined inner hub in response to the shift fork, forcing the blocking ring against the gear cone and then, when the gear is at the same speed, slides over the blocking ring and gear engagement teeth, locking the gear to the synchronizer hub and shaft.

sleeving - a means of reconditioning an engine by boring the cylinder oversize and installing a thin metal liner called a sleeve.

slip - condition caused when a driving part rotates faster than a driven part.

spark knock - engine noise caused by abnormal, uncontrolled combustion due to preignition or detonation.

spark plug - an electrical device that is connected to a high voltage source. The high voltage travels down an electrode inside the spark plug and arcs across an air gap, thereby creating the spark that starts the combustion process in the combustion chamber.

specific gravity - the ratio of the weight or mass of the given volume of a substance to that of an equal volume of another substance, e.g. - water for liquids and solids; air or hydrogen for gases, are used as standards.

splines - external or internal teeth cut into a shaft; splines are used to keep a pulley or hub secured on a rotating shaft.

split-ball gauge - a transfer measuring instrument. Turning the handle on the gauge causes the split ball to expand. It can be used for measuring small holes such as valve guides.

split locks - (see *valve keepers*).

stoichiometric - chemically correct. An air/fuel mixture is considered stoichiometric when it is neither too rich nor too lean; an ideal mixture is composed of 14.7 parts air to one part fuel.

straightedge - a long, flat steel strip with perfectly straight edges, used for checking surfaces for warpage.

stress - force or strain to which material is subjected.

Glossary of Terms

stroke - the distance the piston moves from TDC to BDC.

supercharger - a compressor, mechanically driven by the engineís crankshaft, that pumps air into the intake manifold.

surging - a condition in which the engine speeds up and slows down even when the throttle is held steady.

swirl combustion - a swirling of the air/fuel mixture in a corkscrew pattern that improves combustion.

--*t*--

tang - a projecting point or prong designed to fit into a handle or shaft; another name for the main or rod bearing shell location lug; a projecting piece of metal placed on the end of the torque converter, used to rotate the oil pump.

tap - to cut threads in a hole with a tapered, fluted, threaded tool; a tool used to cut threads in a hole or bore

taper - a gradual decrease in width or thickness; the difference in diameter between the cylinder bore at the bottom of the hole and the bore at the top of the hole, just below the ridge.

TDC - (see *top dead center*).

tension - stress exerted on an object by pulling that tends to extend the material.

tensioner - a device used with a timing chain or belt to maintain its tension.

thermal cleaning - a parts cleaning method that uses high temperature in a bake oven to turn grease, oil and sludge into a powdery residue. This residue is then removed by washing, airless shot blasting or glass beading.

thermal efficiency - ratio of work accomplished compared to total quantity of heat contained in fuel. Fuel contains potential energy in the form of heat when burned in the combustion chamber.

thermostat - a device installed in the cooling system that allows the engine to come to operating temperature quickly and then maintain a minimum operating temperature.

thread chaser - a tool for cleaning threads that will not remove any metal.

thread pitch - the number of threads in one inch of threaded bolt length. In the metric system, the distance in millimeters between two adjacent threads.

throw - distance from center of the crankshaft main bearing to center of the connecting rod journal.

thrust load - load placed on a part that is parallel to the center of the axis.

thrust plate - a plate behind the cam sprocket that controls camshaft end thrust.

timing - refers in crankshaft degrees to the position of the piston in the cylinder. When referring to camshaft timing, it is when the valves open. When referring to ignition timing, it is when the spark occurs.

timing belt - a toothed reinforced belt used to drive the camshaft from a sprocket on the crankshaft.

timing chain - a chain that drives the camshaft from a sprocket on the crankshaft.

timing gears - gears that drive the camshaft from the crankshaft.

tolerance - the difference between the allowable maximum and minimum dimensions of a mechanical part; the basis for determining the accuracy of a fitting.

top dead center (TDC) - the position of the crankshaft for a specific cylinder when the piston is at the highest point in its vertical travel on the compression stroke.

torque - twisting effort on a shaft or bolt.

torque sequence - a specified order in which a componentís mounting bolts should be tightened.

torque-to-yield head bolts - head bolts that are often not reusable. They are torqued into yield, which means that they have purposely stretched beyond the point where they will return to their original length. This provides more uniform clamping force.

torque-turn - the method used to tighten torque-to-yield head bolts. A torque angle gauge is used to tighten a fastener a specified number of degrees after it is torqued to a foot pound specification.

torque wrench - a breaker bar or ratchet wrench with an indicator that measure the twisting effort applied to a fastener during tightening.

transverse - perpendicular or at a right angle to a front-to-back centerline.

turbocharger - an exhaust driven pump which compresses intake air and forces it into the combustion chambers at higher than atmospheric pressure. The increased air pressure allows more fuel to be burned and results in increased horsepower being produced.

--u--

umbrella type valve seals - valve guide seals that fit tightly on the valve stem. They move up and down with the valve stem acting like an umbrella to shield oil away.

undersize - when an inside or outside diameter has been machined to a dimension smaller than standard. Undersized bearings are used to compensate.

--v--

vacuum - a pressure lower than atmospheric.

vacuum advance - a distributor mounted mechanism that controls spark advance in response to engine vacuum.

vacuum gauge - an instrument used to measure the amount of vacuum produced by an engine.

valve - a device that controls the pressure, direction or rate of flow of a liquid or gas.

valve cover - an enclosure fastened to the top of a cylinder head, over the valvetrain.

valve duration - the length of time, in degrees of crankshaft rotation, that a valve is open.

valve face - the area of the valve that contacts the valve seat.

valve float - when valves remain open, usually at high rpm, due to weak or broken valve springs.

valve guide - a bore in the cylinder head that the valve stem fits into.

valve guide knurling - a method of refinishing the inside of a valve guide by restoring its original size.

valve guide liner - a thin bronze bushing installed in a valve guide to restore it to original size.

valve keepers - small locks that hold the valve retainer onto the valve stem. Also called split locks.

valve lash - the amount of clearance in the valvetrain when the lifter is on the base circle of the camshaft lobe.

valve lift - the distance from the valve seat when the valve is fully open.

valve lifter - a small cylinder that fits into a bore above the cam lobe. It acts on a pushrod and rocker arm to open the valve.

valve margin - on a poppet valve, the space or rim between the surface of the head and the surface of the valve face.

valve rotator - a part found at the end of some valve springs that rotates the valve each time it opens. This aids in providing even cooling to the valve.

valve seal - a seal located over the valve stem, used to prevent oil from leaking down the valve guide and into the combustion chamber.

valve seat - the machined surface that the valve face seats against.

valve spring - a small coil spring used to keep the valve closed against the valve seat.

valve spring compressor - a tool used to compress the valve spring on a cylinder head. When the valve spring is compressed, the valve keepers can be removed, then the spring is released and the spring, valve spring retainer and valve can be removed from the cylinder head.

Glossary of Terms

valve spring installed height - the specified distance between the machined spring seat on the cylinder head to the underside of the valve spring retainer. Both grinding the valve and grinding the valve seat result in an increase in this dimension. A shim can be installed under the spring to restore the original installed height for proper spring tension.

valve spring retainer - the part that connects the valve spring to the valve and holds the valve against the cylinder head. It is held to the valve by keepers.

valve spring seats - metal shims used, usually on aluminum cylinder heads, to protect the head from the bottom of the valve spring.

valve stem - the part of the valve that is inside the valve guide.

valve timing - set rotations of the camshaft and crankshaft to open/close valves at proper intervals during the piston strokes for optimal operation of an engine.

valvetrain - parts that convert camshaft movement to valve movement. These include the camshaft, cam timing parts, lifters or cam followers, pushrods, rocker arms, valve and spring.

valvetrain geometry - the dynamic relationship between the rocker arm and valve stem during the time when the valve is opening and closing.

vibration damper - (see ***harmonic balancer***).

viscosity - the rating of a liquid's internal resistance to flow.

volt - unit of electromotive force. One volt of electromotive force applied steadily to a conductor of one-ohm resistance produces a current of one ampere.

voltage drop - voltage lost by the passage of electrical current through resistance.

voltmeter - a tool used to measure the voltage available at any point in an electrical system.

--w--

warpage - a condition that exists when a part is bent or twisted; the degree to which a part deviates from flatness.

waste gate - a bypass valve that limits boost produced by a turbocharger.

water jacket - also called a cooling jacket, it is the hollow area of a casting designed for coolant flow.

water pump - device used to circulate coolant through the engine.

wet sleeve - a sleeve, which when installed in a cylinder block, is exposed to coolant.

wrist pin - a hollow metal tube that secures the piston to the connecting rod and allows the piston to swivel on the rod. Also called a piston pin.

NOTES

NOTES